Engineering Applications of Computational Methods

Volume 4

Series Editors

Liang Gao, State Key Laboratory of Digital Manufacturing Equipment and Technology, Huazhong University of Science and Technology, Wuhan, Hubei, China

Akhil Garg, School of Mechanical Science and Engineering, Huazhong University of Science and Technology, Wuhan, Hubei, China

The book series Engineering Applications of Computational Methods addresses the numerous applications of mathematical theory and latest computational or numerical methods in various fields of engineering. It emphasizes the practical application of these methods, with possible aspects in programming. New and developing computational methods using big data, machine learning and AI are discussed in this book series, and could be applied to engineering fields, such as manufacturing, industrial engineering, control engineering, civil engineering, energy engineering and material engineering.

The book series Engineering Applications of Computational Methods aims to introduce important computational methods adopted in different engineering projects to researchers and engineers. The individual book volumes in the series are thematic. The goal of each volume is to give readers a comprehensive overview of how the computational methods in a certain engineering area can be used. As a collection, the series provides valuable resources to a wide audience in academia, the engineering research community, industry and anyone else who are looking to expand their knowledge of computational methods.

More information about this series at http://www.springer.com/series/16380

Xuewen Wang · Bo Li · Rui Xia · Haozhou Ma

Engineering Applications of Discrete Element Method

Operation Analysis and Optimization Design of Coal and Agricultural Machinery

Springer

Xuewen Wang
College of Mechanical and Vehicle Engineering
Taiyuan University of Technology
Taiyuan, Shanxi, China

Bo Li
College of Mechanical and Vehicle Engineering
Taiyuan University of Technology
Taiyuan, Shanxi, China

Rui Xia
College of Mechanical and Vehicle Engineering
Taiyuan University of Technology
Taiyuan, Shanxi, China

Haozhou Ma
College of Mechanical and Vehicle Engineering
Taiyuan University of Technology
Taiyuan, Shanxi, China

ISSN 2662-3366 ISSN 2662-3374 (electronic)
Engineering Applications of Computational Methods
ISBN 978-981-15-7976-9 ISBN 978-981-15-7977-6 (eBook)
https://doi.org/10.1007/978-981-15-7977-6

This Springer imprint is published by the registered company Springer Nature Singapore Pte Ltd.
The registered company address is: 152 Beach Road, #21-01/04 Gateway East, Singapore 189721, Singapore

Preface

Bulk materials refer to a large number of unpacked bulk, granular and powdery materials which are piled together and whose geometric size basically belongs to the aggregate of particles of the same magnitude, such as rock, sand, coal, grain, cement and so on widely existing in nature. The physical properties of bulk materials are very complex and contain some properties of both solids and fluids. Firstly, the bulk material itself can maintain its solid shape; to a certain extent, it can bear shear and pressure; and generally cannot withstand tensile forces. Secondly, when the external force is applied or the internal force changes, it will guide the flow of bulk materials, but the difference between it and pure fluid is that it cannot transfer the same pressure in all directions. In addition, the characteristics of stress distribution such as arching phenomenon, grain silo effect, Brazil nut effect and self-organized collapse phenomenon make the bulk materials also have special features different from solids and fluids. Current numerical methods for continuous media, such as finite element and boundary element methods, cannot well simulate the bulk system because they all require the continuity of the problem to be studied and the stress balance and displacement coordination conditions must be satisfied.

As a numerical method for discontinuous media, discrete element method (DEM) does not require continuity of the problem studied. Simple contact constitutive relationship and Newton's law of motion are used for simulation calculation, which can be used to solve and analyze the motion and mechanical properties of complex discrete systems. With the continuous development of DEM in recent years, it has been widely used in the fields of industry and agriculture. It can better simulate mechanical and kinetic problems such as coal mining, coal transportation, soil excavation, crop transportation, and can also be used to simulate and analyze the force problems of geotechnical engineering, crop and related machinery.

This book introduces the engineering application of DEM, especially the simulation analysis of the typical equipment (scraper conveyor, coal silos and subsoiler) in the coal and agricultural machinery. In this book, DEM was applied to build rigid-discrete coupling model, and the kinematic effect of the bulk materials, the mechanical effect of the interaction between the bulk materials and the mechanical equipment in the operation process of the relevant equipment were studied. On this basis, the optimization design strategy of the relevant structure was proposed. This book can effectively promote the application of DEM in engineering, analyze the operation state, failure mechanism and operation effect of related equipment in operation, and provide the theoretical bases of the optimal design of equipment. The book is intended for undergraduate and graduate students who are interested in mechanical engineering, researchers investigating coal and agricultural machinery, and engineers working on designing-related equipment.

The book consists of 10 chapters, of which Chaps. 1, 6, 7 and 10 were written by Xuewen Wang, Chaps. 8 and 9 by Bo Li, Chaps. 4 and 5 by Rui Xia, and Chaps. 2 and 3 by Haozhou Ma. The research results and publication of this book are supported by projects such as the National Natural Science Foundation of China: 'The coupling effect between coal and rock bulk and transport equipment under complex transport conditions' (51875386), and the National Natural Science Foundation of China: 'Design of wear-resistant bionic middle groove of scraper conveyor under variable factors based on discrete element method' (51804207).

Due to the author's limited knowledge level, there are some defects in the book, so we sincerely ask the readers to criticize and correct them.

Taiyuan, China
June 2020

Xuewen Wang
Bo Li
Rui Xia
Haozhou Ma

Contents

Chapter 1
Introduction

1.1 Brief Introduction of DEM

The discrete element method (DEM) was first proposed by Professor Cundall based on molecular dynamics in 1971, and was first applied to the analysis of related problems in rock mechanics. The basic principle is to divide the research object into a finite number of particle units, and each unit has a relatively independent motion. Then according to Newton's second law and the interaction between particle units, the dynamic relaxation method or static relaxation method is adopted to carry out cyclic iterative calculations to obtain the force and displacement of all units in each time step, and then update the positions of all units. Then by tracking the microscopic motion of each unit, the macroscopic motion law of the whole research object is obtained.

DEM simulates the motion of each particle in the particle system. In the beginning, DEM regarded particle units as two-dimensional rigid polygonal blocks, and then developed into two-dimensional disks and three-dimensional spheres. Because the size and shape of the disk or sphere are easy to define, and the contact state of particles is easy to detect, the simulation time is short. However, in the actual production, the actual shape of most particles is very different, and the disk and sphere cannot truly reflect the shape of particles, so there are some non-circular particle models such as spheroid, ellipsoid, polyhedron and so on. The non-circular particle models can truly reflect the real shape of the particles, but the complex geometries of these shapes also bring difficulties to the contact judgment of the discrete element solution, and also affect the calculation time. At present, more particle models are still represented by disks, spheres or clusters. The interaction between units is regarded as a transient equilibrium problem by DEM. When the internal force reaches equilibrium, it can be considered to be in equilibrium. And the force on the unit only depends on itself and other units in direct contact with the unit.

X. Wang et al., *Engineering Applications of Discrete Element Method*, Engineering Applications of Computational Methods 4,
https://doi.org/10.1007/978-981-15-7977-6_1

The solution process is to first discrete the solution area into discrete unit arrays, then select the contact model which is consistent with the actual situation. According to the relationship between the force and the relative displacement of the adjacent unit, the normal force and the tangential force of the particle unit are obtained, and then the resultant force and the resultant moment of all the external forces on the units are calculated. The acceleration is obtained by Newton's law of motion. And then the acceleration obtained is integrated with the time to obtain the velocity and displacement of the unit. In this way, the physical quantities of displacement, velocity, acceleration and angular velocity of all units at any time can be obtained by a large number of iterative calculations.

1.2 The Development History of DEM

Since DEM was proposed, it has attracted many scholars to study it, so that its models and algorithms have been continuously improved, and its application field has been further expanded.

In 1971, Cundall first proposed DEM to solve the mechanical behavior of discontinuities such as rocks. The two-dimensional angle-edge (surface) contact model proposed by Cundall can effectively describe the discontinuities of rocks and their particle dispersion. It allows the units to change the original contact relationship and can study the contact conditions with deformation and displacement [1, 2]. In 1974, Cundall studied the input and output of interactive computer in discrete element program and then studied the interactive conversation function of screen graphics output. The two-dimensional discrete element program tended to mature [3–5]. But initially due to the limitations of computer hardware, many programs were written in assembly language. In 1978, the program of DEM had been translated into the text of FORTRAN IV and became the basic program of DEM [6]. In the same year, Cundall and Strack developed the BALL program of two-dimensional disk, and then developed the TRUBAL program of three-dimensional sphere to study the mechanical behavior of particle media [7, 8]. The achievements of Cundall and Strack were published in 'Geotechnique' in 1979, which aroused widespread concern and high attention in the academic community, and officially used the academic term 'Discrete Element Method'. It proved that DEM is a useful way to study the constitutive relationship of particles [9].

In 1977, Kawai et al. proposed a new discrete model in the solid mechanical model. The rigid units in this model are connected by spring units, which can be used to analyze the elastic–plastic stress field near the crack under two-dimensional and three-dimensional arbitrary boundary conditions [10]. In 1980, Cundall developed the Universal Distinct Element Code (UDEC) program based on the original rigid body model, and studied the stress deformation of the block and the block units allowed to break according to the fracture criterion, which made the DEM of rigid block develop into the DEM of variable block [11]. In 1983, Dowding et al. proposed an edge–edge (face–face) contact model based on the angle-edge (face) contact model. This model allows units to be separated without the need for element contact relationships to be maintained, and it does not need to retrieve the contact relationship of the unit, which not only greatly improves the calculation efficiency but also avoids the 'locked state' of the angular-edge contact model. At the same time, a two-dimensional discrete element program NURBM-2D (Northwestern University Rigid Block Model 2 Dimension) was developed based on the rigid block theory, which is used to analyze the stability of underground rock mass projects such as caves and tunnels, and the movement under seismic load [12–14]. In 1985, Shi established a new model for analyzing the static and dynamic characteristics of the block system, and evolved the solution of DEM to discontinuous deformation analysis [15, 16]. In 1987, Cundall analyzed the relationship between the discontinuous medium properties and physical properties of rock, and developed a three-dimensional discrete element program 3DEC for rigid and variable units with ITASCA company. Compared with UDEC, its data structure has been greatly improved. In 1988, Gilbert proposed the three-dimensional discrete element basic equations and algorithms for the face–face contact model [14–17], and developed the FORTRAN text three-dimensional program NURBM-3D. In 1989, Thornton developed the TRUBAL program from the particle contact model to form TRUBAL-Aston (later named GRANULE), which conforms to the principle of elastic-plastic ball contact mechanics and can simulate dry–wet, elastic–plastic and particle two-phase flow problems. Thornton et al. [18–22] fully adopted the previous theory of normal/tangential action in sphere contact mechanics, including the case of surface adhesion and local plastic deformation in contact area. For non-adhesive spherical particles, the Hertz theory is used to solve the normal contact force and the Mindlin-Deresiewicz method to solve the tangential contact force. For adhering spherical particles, the normal contact force is based on the JKR theory that considers the adhesion force based on Hertz theory, and the increment of tangential contact force is determined by Thornton method which is formed by the combination of Savkoor and Briggs method [23] and Mindlin-Deresiewicz method.

Tsuji proposed the nonlinear Hertz contact model based on the Cundall model, obtained the damping form of the contact model, and obtained the expression of the contact force and contact displacement that more reflects the contact law [24, 25]. Oda et al. proposed a simplified form of contact moment, considering the asymmetry of normal contact stress when particles contact based on the relevant experimental study of bulk mechanics [26–28]. This model significantly reduces the error of the calculated results due to the rolling property of the ideal circular particles, and further studies the shear effect and torque transfer of the particles [29, 30]. Lu et al. proposed a contact relationship detection algorithm for the problem of 'locked state' of angular–angular contact [31, 32].

1.3 Basic Theory of DEM

1.3.1 Contact Theory of DEM

1.3.1.1 Particle Model Hypothesis

In order to facilitate the analysis and research, the following assumptions are made for the particle model when the DEM is used for calculation and solution:

(1) The particle unit is a rigid body, and the deformation of the entire system is the sum of all particles deformed at the contact point.
(2) The contact between particles is point contact.
(3) The particle contact characteristic is soft contact, that is, the particles are allowed to produce a certain amount of overlap at the contact point, the resulting overlap is small relative to the particle size, and the particle deformation is also small relative to the particle translation and rotation.
(4) In each time step, the disturbance can only be transmitted to the particles themselves, but not to other particles. At all times, the resultant force on each particle can only be determined by the interaction of other particles in contact with that particle.

1.3.1.2 Particle Simplified Model

At present, DEM often uses two particle simplified models, soft sphere model and hard sphere model. These two models have different advantages in computing efficiency and application. The soft sphere model is mainly used to simulate the collision process between multiple particles over a period of time. According to the overlap between the particles, the contact force between the particles can be calculated by Newton's second law. The calculation intensity is relatively small, which is suitable for numerical calculation of engineering problems. The hard sphere model is mainly used to simulate the case of fast particle velocity, such as coulter flow and shear flow, in which the particles collide instantaneously without obvious plastic deformation. This model only considers the collision of two particles at the same time, and cannot be used to calculate the collision among three or more particles.

(1) Soft sphere model

As shown in Fig. 1.1, the soft sphere model simplifies the process of contact and collision between the particles into the damped vibration of spring vibrator, and its motion equation is:

$$m\ddot{x} + c\dot{x} + kx = 0 \tag{1.1}$$

where x is the displacement of the vibrator from its equilibrium position; m is its mass; c and k are the spring damping coefficient and elastic coefficient, respectively.

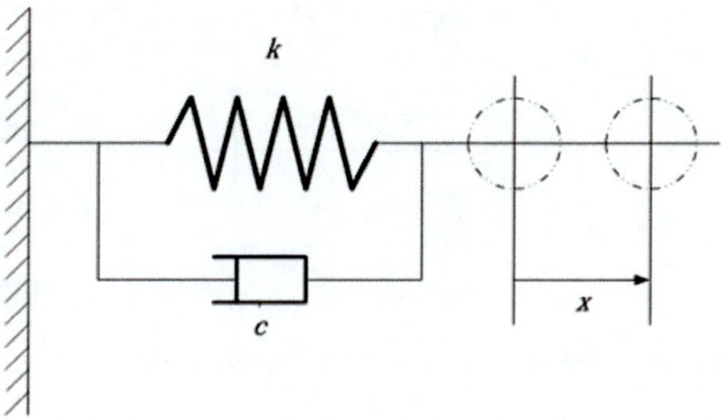

Fig. 1.1 Spring damper system

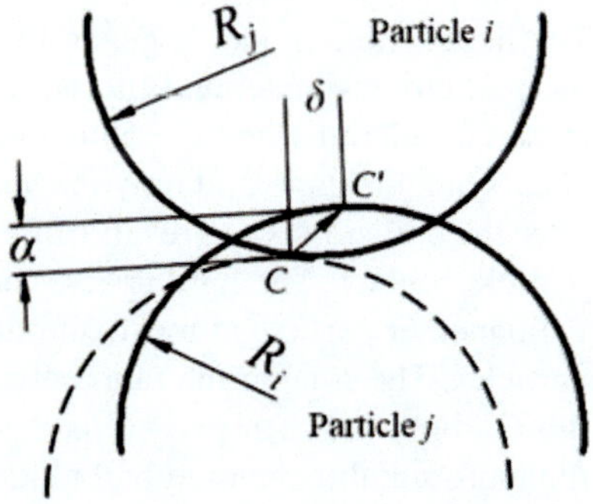

Fig. 1.2 A soft sphere model of two particles

According to formula (1.1), the particles are subjected to the restoring force in the direction opposite to the displacement, and the viscous resistance in the opposite direction, which is proportional to the speed, so the energy of this system is gradually declining.

As shown in Fig. 1.2, particle i and particle j contact at point C due to inertia or external force, and the position at the beginning of contact is indicated by the dotted line. The particle surface will be deformed due to the relative motion of the two particles and generate contact force. The soft sphere model obtains the contact force between the particles by calculating the normal overlap amount α and the tangent displacement δ, without considering the deformation.

The soft sphere model simplifies the contact between particle i and particle j. The simplified model is shown in Fig. 1.3. Springs, couplers, dampers and sliders are set between the two particles, and relevant parameters are introduced. The coupler is only used to determine the pairing relationship between the two particles in contact without introducing any force. In the tangential direction, if the tangential force exceeds a certain value, the sliding resistance device will realize two particles sliding under the action of friction and normal force.

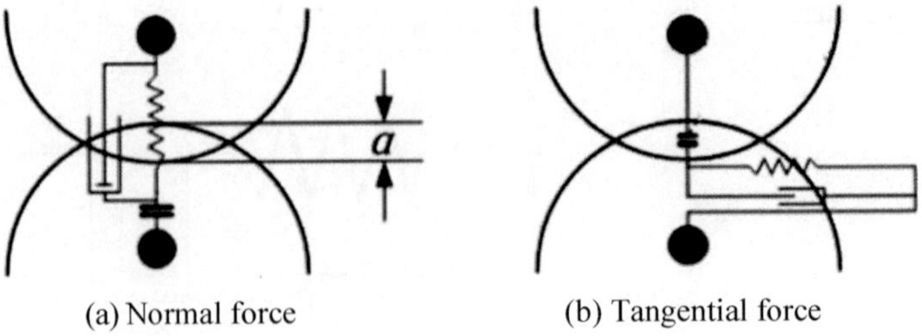

Fig. 1.3 Simplified model of the soft sphere model

(2) Hard sphere model

The hard sphere model is generally used when the particle movement is violent or the particle is sparse, and the particle deformation and contact force do not need to be considered when the collision occurs. It can be considered that the collision occurs at an instant, and only the speed after the collision is determined. The speed after the collision is the result of the integration of force and time in the process of particle contact. The hard sphere model assumes that at any time, particle collides with another particle at most, and the collision point is the contact point of the two particles. The smaller the ratio between the duration of particle contact and the free motion time between two collisions of particles, the sparser the particulate matter, the more suitable the hard ball model.

1.3.1.3 Particle Micro-Parameters

DEM regards the bulk material as an independent, contacting and interacting particle unit and assigns real material properties to the unit by parameters. The micro-parameters of particle system can be divided into three categories: intrinsic parameters of bulk material, contact parameters between particles and between particles and geometry, and preset parameters of contact model. Intrinsic parameters and contact parameters are necessary parameters for discrete element simulation (Table 1.1).

(1) Intrinsic parameters of bulk materials. These parameters are the material's own characteristic parameters. They are relatively fixed and are generally not affected by other factors.
(2) Contact parameters of bulk materials. These parameters are physical parameters of interaction between two objects when they are in contact. They are affected by two objects in contact with each other and the contact parameters between different materials vary greatly. The contact parameters between the same materials will vary depending on the external conditions and the purpose of the study.
(3) The preset parameters of the contact model. Some contact models require preset parameters, such as bonding contact model which can be used to simulate

Table 1.1 Main micro-parameters of particle system

Category	Name
Intrinsic parameters	Poisson's ratio (ν)
	Density (ρ)/kg/m^3
	Shear modulus (G)/Pa
Contact parameters between particles	Coefficient of restitution
	Coefficient of static friction
	Coefficient of rolling friction
Contact parameters between particles and geometry	Coefficient of restitution
	Coefficient of static friction
	Coefficient of rolling friction

bulk breakage needs to set five parameters: normal contact stiffness, tangential contact stiffness, normal critical stress, tangential critical stress and bond radius. JKR cohesion contact model which can characterize bulk water content needs to set JKR surface energy.

1.3.1.4 Contact Model

As one of the necessary components of DEM, the contact model is essentially an elastoplastic analysis of the contact mechanics of granular solids under quasi-static conditions. The force and torque on the particles can be obtained through analysis and calculation. In order to adapt to different research objects, DEM has developed various contact models.

(1) Hertz-Mindlin (no slip) contact model

Hertz-Mindlin (no slip) contact model is the basic contact model for discrete element simulation. The normal force component of the model is based on Hertz theory [33] and the tangential force component is based on the research work of Mindlin-Deresiewicz [34].

Assume that two spherical particles with radii R_1 and R_2 are in elastic contact. α is the amount of normal overlap, and its calculation formula is:

$$\alpha = R_1 + R_2 - |\boldsymbol{r}_1 - \boldsymbol{r}_2| \tag{1.2}$$

where $\boldsymbol{r}_1$ and $\boldsymbol{r}_2$ are the position vectors of the two centroids of particles.

The contact surface between particles is circular and the contact radius a is:

$$a = \sqrt{\alpha R^*} \tag{1.3}$$

where R^* is the equivalent particle radius, which can be obtained from the following formula:

$$\frac{1}{R^*} = \frac{1}{R_1} + \frac{1}{R_2} \tag{1.4}$$

The normal force F_n between particles can be obtained from the following formula:

$$F_n = \frac{4}{3}E^*(R^*)^{\frac{1}{2}}\alpha^{\frac{3}{2}} \tag{1.5}$$

where E^* is the equivalent modulus of elasticity, which can be obtained from the following formula:

$$\frac{1}{E^*} = \frac{1-\nu_1^2}{E_1} + \frac{1-\nu_2^2}{E_2} \tag{1.6}$$

where E_1, ν_1 and E_2, ν_2 are the elastic modulus and Poisson's ratio of particle 1 and particle 2, respectively.

The normal damping force F_n^d can be obtained from the following formula:

$$F_n^d = -2\sqrt{\frac{5}{6}}\beta\sqrt{S_n m^*}v_n^{rel} \tag{1.7}$$

where m^* is the equivalent mass, which can be obtained from the following formula:

$$m^* = \frac{m_1 m_2}{m_1 + m_2} \tag{1.8}$$

If the velocity of two particles before collision is v_1 and v_2, respectively, and the normal unit vector at the time of collision is $\boldsymbol{n}$, then,

$$\boldsymbol{n} = \frac{\boldsymbol{r}_1 - \boldsymbol{r}_2}{|\boldsymbol{r}_1 - \boldsymbol{r}_2|} \tag{1.9}$$

V_n^{rel} in the formula is the normal component value of relative velocity, which is obtained from the following formula:

$$v_n^{rel} = (v_1 - v_2)\cdot \boldsymbol{n} \tag{1.10}$$

where the normal stiffness S_n and coefficient β are, respectively, obtained from the following two formulas:

$$\beta = \frac{\ln e}{\sqrt{\ln^2 e + \pi^2}} \tag{1.11}$$

$$S_n = 2E^*\sqrt{R^*\alpha} \tag{1.12}$$

where e is the recovery coefficient.

The tangential force F_t between particles can be obtained from the following formula:

$$F_t = -S_t \delta \tag{1.13}$$

where δ is the tangent repetition and S_t is the tangent stiffness. The tangential stiffness S_t can be obtained from the following formula:

$$S_t = 8G^* \sqrt{R^* \alpha} \tag{1.14}$$

where G^* is the equivalent shear modulus, which can be obtained from the following formula:

$$G^* = \frac{2 - v_1^2}{G_1} + \frac{2 - v_2^2}{G_2} \tag{1.15}$$

where G_1 and G_2 are the shear modulus of particles 1 and 2, respectively.

The tangential damping force F_{td} between particles can be obtained from the following formula:

$$F_{td} = -2\sqrt{\frac{5}{6}} \beta \sqrt{S_t m^*} v_t^{rel} \tag{1.16}$$

where V_t^{rel} is the tangential relative velocity.

The tangential force is related to friction $\mu_s F_n$, where μ_s is the static friction factor.

The rolling friction in simulation is very important, which can be explained by the moment on the contact surface.

$$T_i = -\mu_r F_n R_i \boldsymbol{\omega}_i \tag{1.17}$$

where μ_r is the rolling friction coefficient; R_i is the distance from the center of mass to the contact point; $\boldsymbol{\omega}_i$ is the unit angular velocity vector of the object at the contact point.

(2) Hertz-Mindlin with Archard wear contact model

Hertz-Mindlin with Archard wear contact model is an extension of the standard Hertz-Mindlin (no slip) model in EDEM. It gives an estimate of the surface wear depth of geometry. This model is based on Archard wear theory [35]. In an ideal state, the material removal volume on the surface of geometry is directly proportional to the friction work of particles when they move on the surface of material.

Archard equation:

$$Q = W F_n d_t \tag{1.18}$$

where Q is the material removal volume, d_t is the tangential movement distance, W is the wear constant, and W can be obtained from the following formula:

$$W = \frac{K}{H} \tag{1.19}$$

where K is the dimensionless constant and H is the surface hardness of the material. W value is input in EDEM, which simplifies the input of parameters in EDEM.

Since the equation predicts the volume of material to be removed, a new depth is given to each unit in EDEM:

$$h = \frac{Q}{A} \tag{1.20}$$

where A is the area of the material removed.

(3) The parallel-bond model

Parallel-bond can reflect the effect of additional substances existing between two contacting particles. These additional substances act parallel between two particles, so it is called parallel-bond. Parallel-bond is similar to a cylindrical, elastic bonding between two particles, which works in conjunction with the point contact of the linear contact stiffness model and can withstand the dual effects of force and moment when established. When the contact force between particles exceeds the contact strength, the bond will break, which is consistent with the effect of liquid bridge force in practice. Parallel-bond can be represented by a series of normal and tangential parallel-bond springs. Parallel-bond is defined by the following five parameters, including normal and tangential parallel-bond stiffness, normal and tangential parallel-bond strength and amplification radius coefficient.

The force and moment of parallel-bond are calculated by the following two formulas:

$$\bar{F}_i^{pb} = \bar{F}_{ni}^{pb} + \bar{F}_{si}^{pb} \tag{1.21}$$

$$\bar{M}_i^{pb} = \bar{M}_{ni}^{pb} + \bar{M}_{si}^{pb} \tag{1.22}$$

where $\bar{F}_{ni}^{pb}$ is normal force component of parallel-bond. $\bar{F}_{si}^{pb}$ is tangential force component of parallel-bond. $\bar{M}_{ni}^{pb}$ is normal moment component of parallel-bond. $\bar{M}_{si}^{pb}$ is tangential moment component of parallel-bond.

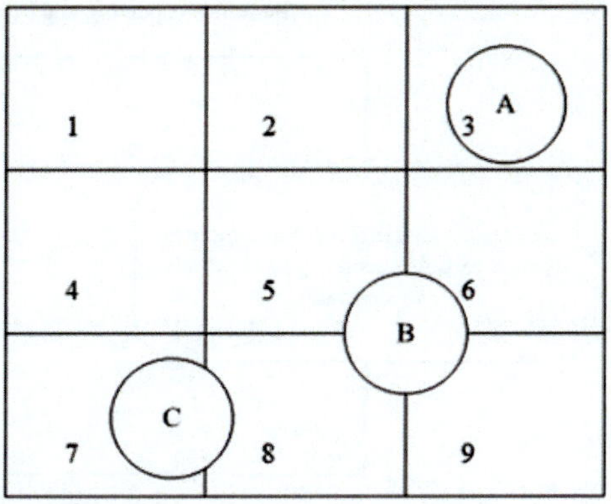

Fig. 1.4 Mesh method for particle contact search

1.3.2 Particle Contact Search

When DEM is used to simulate the bulk materials, it is necessary to determine which particles are in contact first, and then calculate the contact force of the particles in contact according to the defined contact model to update the speed and position of each particle. The search of particle contact is a prerequisite for calculating the force on particles. In the simulation, the search of particle contact should be carried out every other time step. In each search, it is very important for the calculation of contact force to locate the particles in contact quickly and accurately. Therefore, it is very important to search the contact between particles efficiently.

The mesh division method is mainly used to search particle contact, that is, the calculation area is divided into several square meshes according to the maximum diameter of particles. The relationship between the mesh side length l_{box} and the maximum diameter $d_{\max}$ is $d_{\max}<l_{\text{box}}<2d_{\max}$.

The basic idea of mesh method is to judge which particles are in the same mesh first, and then judge whether the particles are in the same mesh contact. As shown in Fig. 1.4, there are three particles in the mesh, particle A is in mesh 3, particle B is in mesh 5, 6, 8, 9 and particle C is in mesh 7, 8. When retrieving whether the particles are in contact, particles B and C are in mesh 8, so they may be in contact, while particles A and B are in different meshes, hence they are not in contact. Further judge whether B and C contact.

1.3.3 Calculation of Particle Motion State

DEM is to track the motion state of the particles through cyclic calculation. The method is to divide the whole simulation process into time steps with equal interval, determine the interaction between particles in each time step, calculate the particle motion state according to Newton's second law and then calculate the particle motion state in the whole process by time step iteration. As shown in Fig. 1.5, each cycle includes the following two calculation processes:

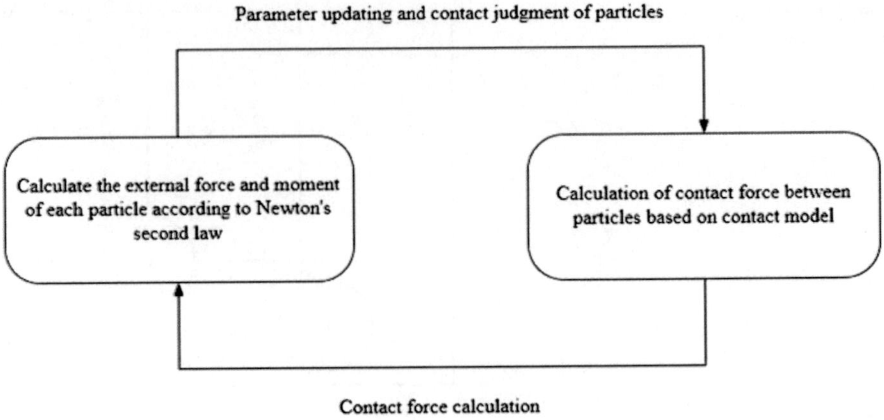

Fig. 1.5 Calculation cycle in iteration time step

(1) Retrieve the position of each particle and determine whether there is contact between the particles, determine the size of the contact force between the particles by the defined contact model and determine the amount of external force received by each particle according to the principle of acting force and reaction force.
(2) According to Newton's second law, the external force on particles and the motion state of particles in the previous time step, the motion state of particles in the next time step is calculated.

DEM is solved by display method, including two sets of equations: motion equation and contact force equation. In the calculation, firstly, the central difference method is used to discredit the motion equation, and an iteration is carried out at each time step to obtain the motion state data such as the particle position, and then the size of the contact force is obtained according to the contact force equation as the initial value of the next iteration to find the new position of the particle, so the iteration is repeated until the agreed end time point of the simulation.

1.3.4 Determination of Time Step

The time step iteration method stipulates that the force on the particles in each time step is invariable, that is, the acceleration is invariable. Therefore, the choice of time step size has a significant impact on the accuracy of simulation results. Too large time step selection will lead to inaccurate description of particle contact process and divergence of numerical calculation; too small time step selection will lead to a large increase of calculation. Now, the Rayleigh wave method is often used to determine the time step. The Rayleigh wave refers to the polarized wave propagating along the particle surface when the particles contact each other and the surface is affected by variable stress. When particles collide, 70% of the total energy consumption is consumed by the Rayleigh wave. Therefore, the critical time step is determined according to the Rayleigh wave velocity along the surface of solid spherical particles. In the process of Rayleigh wave propagation, the energy on the particle surface is the strongest, and it weakens significantly with the increase of depth. Generally, when the propagation depth is more than twice the wavelength, the amplitude is very small.

1.4 Development Trend of DEM

1.4.1 Development of Theoretical Research

After decades of development, DEM has made great progress in various related theories. However, due to its inherent shortcomings of theoretical rigor, there are assumptions in motion, stress, deformation and other aspects. Under these assumptions, the simulation results may deviate from the reality greatly. Therefore, how to define and set the relevant parameters of the discrete element reasonably, so that it can reflect the real particle system as much as possible, these need to be improved in theory. Strengthen the in-depth study of the discrete element theory, maintain its advantages in the study of bulk materials, and effectively describe the stress state and deformation state of the local continuous medium so that the DEM can establish a model that truly meets the geometric conditions, physical conditions and stress conditions. At the same time, strengthen the analysis and comparison of theoretical results, experimental results and numerical simulation results, find out the shortcomings of the discrete element theory and improve it to improve the reliability of numerical simulation.

1.4.2 Improvement of Operation Efficiency

When using DEM for numerical simulation, scholars often encounter irregular particle geometry, particle system composed of a large number of particles and so on. Complex data structure, block retrieval, contact detection and so on will take up a lot of memory and computing time. Under the existing conditions, a group of simulation often takes weeks or even a month. How to effectively calculate and improve the efficiency of calculation on the premise of ensuring the accuracy of calculation is an urgent problem to be solved. In addition to continue to improve the theory of discrete element, scholars can also develop relevant algorithms to solve the problem of operational efficiency.

1.4.3 Multi-field Application

DEM was originally proposed to solve the problems related to rock mechanics. With the continuous development of many scholars, DEM has been widely used in geotechnical engineering, agricultural engineering, mining engineering, mechanical engineering and other fields. It can better simulate mechanical problems and motion problems such as coal and rock mining, coal and rock transportation, soil excavation, crop transportation, and can also be used to simulate and analyze the stress problems of geotechnical, crop and related machinery. DEM can analyze the operation problems caused by particle size, reduce the demand for physical prototype and test, obtain the information of particle size behavior that is not easy to measure, and determine the impact of particle flow on fluid behavior or machinery. With its continuous development, the application fields of DEM will continue to expand. At present, it has been able to solve the problems of mixing and separation of bulk materials, shrinkage, fracture and agglomeration, particle damage and wear, solid–liquid flow conditions, mechanical response of machine components to particle collision, corrosion, particle packaging and surface treatment, heat and mass transfer, chemical reaction dynamics, sedimentation and particle migration removal of solid–liquid system, treatment of hazardous materials, compression of dry and wet solids, viscosity and plastic mechanics, behavior of colloids and glasses, and so on.

1.4.4 Software Development

ITASCA company has developed and improved the discrete element program based on two-dimensional disk shape and three-dimensional sphere, and developed it into PFC^{2D} and PFC^{3D}. Because of its advantages of fast search of adjacent elements and effective simulation of large deformation, the DEM has made a big step forward in engineering application. With the continuous development of science and technology, British DEM-Solutions company has launched the first general CAE software EDEM designed with modern discrete element model technology to simulate and analyze particle handling and production operations. The mechanical analysis series software GDEM jointly developed by the Joint Laboratory of Discontinuous Media Mechanics and Engineering Disaster of the Chinese Academy of Sciences and Beijing JidaoChengran Technology Co., Ltd. realizes the continuous and discontinuous unified calculation. There are also Rock DEM, MatDEM and other software, as well as YADE and other open-source discrete element software, which constantly enrich and develop the engineering application of DEM, and expand the application field of DEM. At present, the main trend of software development is to simplify user operation, establish a friendly interface, simulate irregular particles and improve the efficiency of operation.

1.4.5 Multi-algorithm Coupling Analysis

With the continuous expansion of the application of DEM, it is difficult to fully reflect the real situation only using the DEM. In order to solve this situation, many scholars coupled the DEM with other algorithms, and then achieved the purpose of analysis.

In order to analyze the research between the bulk materials and the machinery, scholars combine the DEM and dynamics, and use the dynamics software to simulate the complex moving machinery to carry out DEM-MBD coupling calculation. The traditional numerical methods, such as finite element method and boundary element method, are suitable for solving the problem of continuous medium, while the DEM is suitable for the problems of discontinuous medium with weak interface connection and the problems of material damage and failure in the transition from continuum to non-continuum. Therefore, in order to give full play to their advantages and expand the scope of application, scholars also carry out the coupling calculation of DEM-FEM, DEM-BEM and so on. With the further development of application field, the coupling analysis of DEM and other algorithms will become an inevitable trend.

1.5 Parameter Calibration of DEM

For the current numerical simulation using discrete element software, accurate and reasonable definition of relevant discrete element micro-parameters will affect the reliability of the final result. Therefore, the parameters of discrete element must be calibrated before numerical simulation, which will directly affect whether the simulation results are true and reliable [37].

1.5.1 Direct Measurement of the Test

The direct measurement of the test is based on the relevant theory and standards to design appropriate tests to measure the particle parameters. Generally, experimental measurement is a quantitative description of material properties, which can better reflect the material properties. At present, the intrinsic parameters of materials are often measured directly by tests. The shear modulus and Poisson's ratio [38, 39] of materials are usually measured by uniaxial compression test and shear test, and the particle density is usually measured by discharge volume method [40].

For the measurement of coefficient of restitution, scholars built an experimental platform based on the principle of free-falling collision (Fig. 1.6a). With the help of high-speed camera or camera, they measured and calculated the relevant physical quantities [41–45], and obtained the coefficient of restitution between the particles and the geometry. The plate was replaced by a plate processed from granular materials [46–51] or a particle plate bonded by particles [52–54], and the coefficient of restitution between particles was obtained using the same test.

For the coefficient of static friction, many scholars [55, 56] designed the inclined plate test based on the traditional lifting inclined plate principle (Fig. 1.6b). Using the lifting mechanism to slowly lift the inclined plate angle, when the particles began to slide, the inclined plate angle was measured, and then the coefficient of static friction between the particles and the geometry was determined. The plate was replaced by

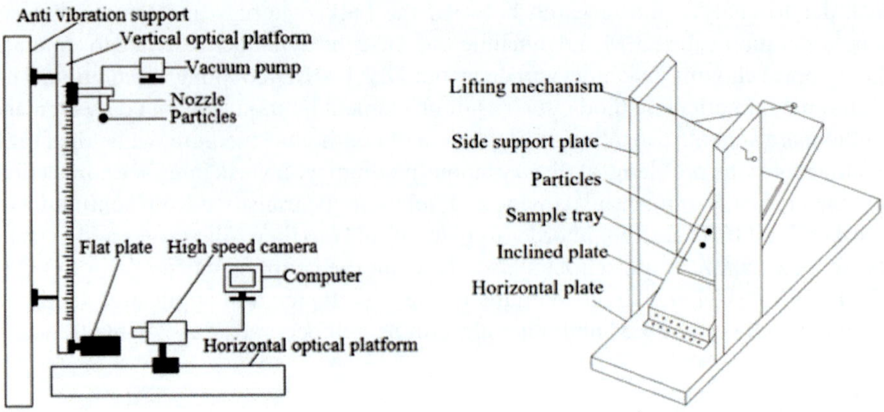

(a) Coefficient of restitution test (b) Inclined plate test

Fig. 1.6 The test of parameter measurement

a plate processed from granular materials or a particle plate bonded by particles [56, 57], and the coefficient of static friction between particles was obtained using the same test. Some scholars also used inclined plate test [58], horizontal traction test [59, 60], rotation test [61–63] and shear test [64, 65] for calculation and measurement.

Many methods have been proposed for the coefficient of rolling friction between the particles and the geometry. Based on the principle of inclined plate lifting test [65, 66], the coefficient of rolling friction is obtained by measuring the angle of inclined plate when the particles begin to roll. However, because the particles are cylindrical and spherical in shape, the slight vibration during the lifting process of inclined plate will make the particles roll, which requires higher test and measurement operation. Some scholars used the inclined plane test to calculate the coefficient of rolling friction of the sample by measuring the rolling distance [67] of the particles on the horizontal plate or the instantaneous speed [68] at two positions.

When the particle system parameters to be calibrated are measured directly, some problems are difficult to solve. Because some parameters of discrete element are affected by other conditions, they cannot meet the relevant conditions of the test accurately. The accuracy of parameter measurement results will be affected by the direct measurement of the test.

1.5.2 Test-Simulation Combined Calibration Method

In order to solve the problem that the parameters of irregular particles cannot be accurately measured by tests, scholars use simulation to measure the parameters of particle system to be calibrated by the test-simulation combined calibration method.

For single-parameter calibration, the macroscopic physical index is determined by single particle test or particle system test; then, the same simulation model as the test is established in the discrete element software, and the corresponding simulation test is carried out to change the value of the parameters to be calibrated to get the macroscopic index under different parameter simulation; then, according to the rule of macroscopic index and parameters to be calibrated, the test is carried out to determine [69–72]. When more than one parameter needs to be calibrated, the calibration process will become complex.

When there are many calibration parameters, to establish the relationship between the test indexes and the parameters to be calibrated quickly and efficiently, scholars proposed many methods to optimize this process, such as test design method [73–75], neural network method based on Latin hypercube sampling [76, 77], artificial neural network method and so on [78]. From the mathematical point of view, the multi-parameter joint calibration is a reverse calibration method [79, 80]. The calibration process can be summarized into three steps: first, determine the evaluation index of the test and the value range of the parameters to be calibrated; second, build the objective function which can reflect the difference between the simulation results and the test results of the simulation model under the combination of the parameters through the discrete element simulation; third, from the actual situation, the optimal combination of parameters [81] satisfying the objective function was found and all parameters to be calibrated were determined.

References

1. Cundall PA (1971) A computer model for simulating progressive largescale movements in blocky system. Muller Led Proc Symp Int Soc Rock Mechanics. Rotterdam, Netherlands: Balkama A A. 1:8–12
2. Cundall, P.A. (1971). The measurement and analysis of acceleration in rock slopes. Dissertation. University of London
3. Cundall PA (1974) Rational design of tunnel supports: A computer model for rock mass behavior using interactive graphics for the input and output of geometrical data. Tech Report MRD-2–74, Missouri River Division, US Army Corps of Engineers
4. Cundall PA (1977) Computer Interactive Graphics and the Distinct Element Method. Rock Engineering for Foundations and Slopes, ASCE. 2:193–199
5. Cundall PA, Voegele MD, Fairhurst C (1977) Computerized Design of Rock Slopes Using Interactive Graphics for the Input and Output of Geometrical Data. Design Methods in Rock Mechanics, ASCE., 1–10
6. Cundall PA, Marti J, Beresford PJ, Last N, Asgian M (1978) Computer modeling of jointed rock masses. Tech Report N-78–4, US Army Engineer Waterways Experiment Station, Vicksburg, Mississippi
7. Strack ODL, Cundall PA (1978) The distinct element method as a tool for research in granular media, Part I. University of Minnesota, Report to the National Science Foundation, Minnesota
8. Cundall PA, Strack ODL (1979) The distinct element method as a tool for research in granular media, Part II. University of Minnesota, Report to the National Science Foundation, Minnesota
9. Cundall PA, Strack ODL (1979) A discrete numerical model for granular assembles. Geotechnique 29(1):47–65
10. Kawai T (1977) New Element Models in Discrete Structural Analysis. J. Soc. Naval Arch. Jpn. 141:187–193

11. Cundall, P.A. (1978). UDEC—A generalized distinct element program for modeling jointed rock. Report PCAR-1-80, Peter Cundall Associates, European Research Office, US Army
12. Dowding CH, Belytschko TB, Yen HJ (1983) Dynamic computational analysis of opening in joint rock. J Geotech Engng. 109(12):1551–1566
13. Belytschko T, Plesha M, Dowding CH (1984) A computer method for the stability analysis of caverns in jointed rock. Int. J. Num. Anal. Meth. Geomech. 22(1):473–492
14. Dowding, C.H., &Gilbert, C. (1988). Dynamic stability of rock slopes and high frequency traveling waves. *J Geotech Engng.*, 114(10), 1069-1 088
15. Shi, G. (1988). Discontinuous Deformation Analysis: a New Numerical Model for the Statics and Dynamics of Block Systems. Dissertation. University of California
16. Shi G (1992) Discontinuous deformation analysis: a new numerical model for the statics and dynamics of deformable block structures. Eng Comput 9(2):157–168
17. Gilbert, C.M. (1988). Development of a three-dimensional small displacement rigid block model for dynamic analysis. Dissertation. Northwestern University
18. Thornton C (1991) Interparticle Sliding in the Presence of Adhesion. J Phys D Appl Phys 24:1942–1946
19. Thornton C, Yin KK (1991) Impact of Elastic Spheres with and without Adhesion. Powder Technol 65:153–166
20. Thornton C (1993) On the Relationship between the Modulus of Particulate Media and Surface Energy of the Constituent Particles. J Phys D Appl Phys 26:1587–1591
21. Thornton C, Ning ZA (1998) Theoretical Model for the Stick/ Bounce Behaviour of Adhesive. Elastic plastic Spheres. Powder Technol. 99:154–162
22. Thornton C (1991) Coefficient of Restitution for Collinear Collisions of Elastic perfectly plastic spheres. J Appl Mech 64:383–386
23. Savkoor AR, Briggs GAD (1977) The Effect of Tangental Force on the Contact of Elastic Solids in Adhesion. Proc R Soc Lond A 356:103–114
24. Tsuji Y, Tanaka T, Ishida T (1992) Lagrangian Numerical Simulation of Plug Flow of Cohesionless Particles in a Horizontal Pipe. Powder Technol 71(3):239–250
25. Tsuji Y, Kawaguchi T, Tanaka T (1993) Discrete Particle Simulation of Two-Dimensional Fluidized Bed. Powder Technol 77(1):79–87
26. Oda M, Konishi J, Nemat-Nasser S (1982) Experimental Micromechanical Evaluation of the Strength of Granular Materials: Effects of Particle Rolling. Mech Mater 1:269–283
27. Oda, M., &Iwashita, K. (1999). Mechanics of Granular Materials, An introduction. Balkema A A, Rotterdam, 147–223
28. Oda, M., Iwashita, K., &Kakiuchi, T. (1997) Importance of Particle Rotation in the Mechanics of Franular Materials. Behringer R P, JenkinsJ T, eds. Powder & Grain 97., 207–214
29. Oda M, Kazama H (1998) Microstructure of Shear Bands and its Relation to the Mechanisms of Dilatancy and Failure of Dense Granular Soils. Geotechnique 48:465–481
30. Oda M, Iwashita K (2000) Study on couple stress and shear band development in granular media based on numerical simulation analyses. Int J Eng Sci 38:1713–1740
31. Lu J, Zhang C, Wang G, Jin F (1996) Three-dimensional distinct element numerical model for static and dynamic analysis of rock bodies. Journal of Tsinghua University (Science and Technology) 36:98–104 (in Chinese)
32. Lu, J. (1996). A study of numerical model of distinct element method and its application in engineering. Dissertation. Tsinghua University
33. Hertz H (1882) On the contact of elastic solids. J Reine Angew Math. 92:156–171
34. Mindlin RD, Deresiewicz H (1953) Elastic Spheres in Contact under Varying Oblique Forces. J Appl Mech 20(3):327–344
35. Archard JF (1953) Contact and rubbing of flat surfaces. J Appl Phys 24:981–988
36. Potyondy DO, Cundall PA (2004) A bonded-particle model for rock. Int J Rock Mech Min Sci 41(8):1329–1364
37. Horabik J, Molenda M (2016) Parameters and contact models for DEM simulations of agricultural granular materials: A review. Biosyst Eng 147:206–225

38. Patwa A, Ambrose RPK, Dogan H, Casada ME (2014) Wheat Mill Stream Properties for Discrete Element Method Modeling. Trans. ASABE. 57(3):891–899
39. Ye Y, Zeng Y (2017) A size-dependent viscoelastic normal contact model for particle collision. Int J Impact Eng 106:120–132
40. Coetzee CJ (2016) Calibration of the discrete element method and the effect of particle shape. Powder Technol 297:50–70
41. Combarros M, Feise HJ, Zetzener H, Kwade A (2014) Segregation of particulate solids: Experiments and DEM simulations. Particuology. 12:25–32
42. Cabiscol R, Finke JH, Kwade A (2017) Calibration and interpretation of DEM parameters for simulations of cylindrical tablets with multi-sphere approach. Powder Technol 327:232–245
43. Horabik J, Beczek M, Mazur R, Parafiniuk P, Ryzak M, Molenda M (2017) Determination of the restitution coefficient of seeds and coefficients of visco-elastic Hertz contact models for DEM simulations. Biosyst Eng 2017(161):106–119
44. Dong M, Mei Y, Li X, Shang Y, Li S (2018) Experimental measurement of the normal coefficient of restitution of micro-particles impacting on plate surface in different humidity. Powder Technol 335:250–257
45. Li X, Dong M, Li S, Shang Y (2019) Experimental and theoretical studies of the relationship between dry and humid normal restitution coefficients. J Aerosol Sci 129:16–27
46. Aryaei A, Hashemnia K, Jafarpur K (2010) Experimental and numerical study of ball size effect on restitution coefficient in low velocity impacts. Int J Impact Eng 37(10):1037–1044
47. Téllez-Medina DI, Byrne E, Fitzpatrick J, Catak M, Cronin K (2010) Relationship between mechanical properties and shape descriptors of granules obtained by fluidized bed wet granulation. Chem Eng J 164(2–3):425–431
48. Gollwitzer, F., Rehberg, I., Kruelle, C.A., &Huang K. (2012). Coefficient of restitution for wet particles. *Phys. Rev. E.*, 86(1)
49. Šibanc R, Kitak T, Govedarica B, Srcic S, Dreu R (2013) Physical properties of pharmaceutical pellets. Chem Eng Sci 86:50–60
50. Sutkar VS, Deen NG, Padding JT, Kuipers JAM, Salikov V, Cruger B, Antonyuk S, Heinrich S (2015) A novel approach to determine wet restitution coefficients through a unified correlation and energy analysis. AIChE J 61(3):769–779
51. Crüger B, Salikov V, Heinrich S, Antonyuk S, Sutkar VS, Deen NG, Kuipers JAM (2016) Coefficient of restitution for particles impacting on wet surfaces: An improved experimental approach. Particuology 25:1–9
52. Li, H. (2012). Research of Modern Design Method for Air-and Screen Cleaning Device. Dissertation. Nanjing Agricultural University
53. Yang, J. (2012). Simulation study of particles flow on the vertical dryer based on DEM. Dissertation. Huazhong Agricultural University
54. Zhang T, Liu F, Zhao M, Ma Q, Wang W, Fan Q, Yan P (2018) Determination of corn stalk contact parameters and calibration of Discrete Element Method simulation. J. China Agric. Univ. 23(4):120–127 (in Chinese)
55. Grima AP, Wypych PW (2011) Development and validation of calibration methods for discrete element modelling. Granul. Matter. 13(2):127–132
56. Han Y, Jia F, Tang Y, Liu Y, Zhang Q (2014) Influence of granular coefficient of rolling friction on accumulation characteristics. Acta Phys. Sin. 63(17):165–171 (in Chinese)
57. Zhou, W. (2015). The Experiment and Analysis of Paddy Field Deep Fertilizing Device Based on Discrete Element Method. Dissertation. Northeast Agricultural University
58. Yang, E. (2015). The experimental research and DEM simulation of the granular material's mixing uniformity in the rotary bowl. Dissertation. Xiangtan University
59. Li Y, Xu Y, Thornton C (2005) A comparison of discrete element simulations and experiments for 'sandpiles' composed of spherical particles. Powder Technol 160(3):219–228
60. Liu, X. (2015). Study and application agricultural material mechanics. Dissertation. Shanxi Agricultural University
61. Barrios GKP, de Carvalho RM, Kwade A, Tavares LM (2013) Contact parameter estimation for DEM simulation of iron ore pellet handling. Powder Technol 248(2):84–93

62. Suzzi D, Toschkoff G, Radl S, Machold D, Fraser SD, Glasser BJ, Khinast JG (2012) DEM simulation of continuous tablet coating: Effects of tablet shape and fill level on inter-tablet coating variability. Chem Eng Sci 69(1):107–121
63. Just S, Toschkoff G, Funke A, Djuric D, Scharrer G, Khinast J, Knop K, Kleinebudde P (2013) Experimental Analysis of Tablet Properties for Discrete Element Modeling of an Active Coating Process. AAPS PharmSciTech 14(1):402–411
64. Frankowski, P, &Morgeneyer, M. (2013). Calibration and Validation of DEM Rolling and Sliding Friction Coefficients in Angle of Repose and Shear Measurements. Aip powders and grains 2013: proceedings of the 7th international conference on micromechanics of granular media, 851–854
65. Ucgul M, Fielke JM, Saunders C (2014) Three-dimensional discrete element modelling of tillage: Determination of a suitable contact model and parameters for a cohesionless soil. Biosyst Eng 121:105–117
66. Budinski KG (2005) An inclined plane test for the breakaway coefficient of rolling friction of rolling element bearings. Wear 259(7):1443–1447
67. Ketterhagen WR, Bharadwaj R, Hancock BC (2010) The coefficient of rolling resistance (CoRR) of some pharmaceutical tablets. Int J Pharm 392(1–2):107–110
68. Zhang, X. (2017). Study on the Drying Process and Equipment for Organic Fertilizer Pellets. Dissertation. China Agricultural University
69. Kim DH, Gratchev I, Berends J, Balasubramaniam A (2015) Calibration of restitution coefficients using rockfall simulations based on 3D photogrammetry model: a case study. Nat Hazards 78(3):1931–1946
70. Grima AP, Wypych PW (2011) Investigation into calibration of discrete element model parameters for scale-up and validation of particle-structure interactions under impact conditions[J]. Powder Technol 212(1):198–209
71. Coetzee CJ, Els DNJ, Dymond GF (2010) Discrete element parameter calibration and the modelling of dragline bucket filling. J. Terramech. 47(1):33–44
72. Wang L, Li R, Wu B, Wu Z, Ding Z (2018) Determination of the coefficient of rolling friction of an irregularly shaped maize particle group using physical experiment and simulations. Particuology 38:185–195
73. Santos KG, Campos AVP, Oliveira OS, Ferreira LV, Francisquetti MC, Barrozo MAS (2015) Dem simulations of dynamic angle of repose of acerola residue: a parametric study using a response surface technique. Blucher Chemical Engineering Proceedings 1(2):11326–11333
74. Wu W, Guo B, Gao Z, Zheng M, Yang H, Li D (2019) Sand modeling and parameter calibration based on DEM. Journal of Chinese Agricultural Mechanization 40(08):182–187 (in Chinese)
75. Xiang W, Wu M, Lv J, Quan W, Ma L, Liu J (2019) Calibration of simulation physical parameters of clay loam based on soil accumulation test. Trans. Chin. Soc. Agric. Eng. 35(12):116–123 (in Chinese)
76. Rackl M, Hanley KJ (2017) A methodical calibration procedure for discrete element models. Powder Technol 307:73–83
77. Zhou H, Hu Z, Chen J, Lv X, Xie N (2018) Calibration of DEM models for irregular particles based on experimental design method and bulk experiments. Powder Technol 332:210–223
78. Benvenuti L, Kloss C, Pirker S (2016) Identification of DEM simulation parameters by Artificial Neural Networks and bulk experiments. Powder Technol 291:456–465
79. Tarantola, A. (2005). Inverse Problem Theory and Methods for Model Parameter Estimation Society for Industrial and Applied Mathematics, Philadelphia
80. Aster, R., Borchers, B., &Thurber, C.H. (2013). Parameter Estimation and Inverse Problem (Second Edition), Elsevier
81. Do HQ, Aragón AM, Schott DL (2018) A calibration framework for discrete element model parameters using genetic algorithms. Adv Powder Technol 29:1393–1403

Chapter 2
Application of DEM in Coal and Agricultural Machinery

2.1 Application of DEM in Coal Machine Equipment

Mining engineering is the basic industry in national economy, which is responsible for providing useful minerals and raw materials for various processing industries. With the continuous development of DEM, its model and algorithm are gradually optimized and improved, which can be well used in mining engineering. Many scholars apply DEM to the research of coal machine equipment.

2.1.1 Transportation Machinery

Many scholars use DEM to explore the problem of transportation machinery of coal mine equipment. Qiu et al. simulated the flow state of ore materials in the middle trough of the scraper conveyor, and studied the material flow characteristics of different structures in the middle trough. The results show that DEM can be well applied to the engineering design of the middle trough [1]. Katterfeld and Gröger applied DEM to simulate the loading and unloading process of bucket elevator and scraper conveyor [2]. Yang and Song used EDEM to establish the simulation model of bulk conveying of belt conveyor, and then carried out the simulation to verify the accuracy of the derived calculation method of material guide groove resistance and material pressure distribution on rollers [3].

Piao et al. simulated the material state at the turn of the belt conveyor by using EDEM, and found that the particle shape and coefficient of rolling friction have little influence on the average velocity of particles at the turn of the belt conveyor. The pressure value of each part of the conveyor belt is mainly affected by the bending degree of the conveyor belt and the speed of conveying particles [4]. Ma and Liu

X. Wang et al., *Engineering Applications of Discrete Element Method*, Engineering Applications of Computational Methods 4,
https://doi.org/10.1007/978-981-15-7977-6_2

applied the DEM to simulate the process of conveying materials by the scraper of the road header, and obtained the dynamic load change of the scraper in contact with the material. The results can be applied to the structural design of the scraper conveyor under impact [5]. Zhou and Wei realized the quantitative analysis of the material distribution of the belt conveyor at each moment through EDEM, and provided a basis for the prediction of the movement state of the conveyed material. At the same time, it proposed that combining with other numerical calculation methods can improve the accuracy of the DEM [6].

Zidek et al. used DEM to simulate the process of conveying materials by a chain scraper conveyor, studied the influence of the scraper geometry on the motion characteristics of the material and further optimized the conveyor model [7]. Cai et al. combined the DEM with the FEM and found the law among the coefficient of static friction between the coal particles and the middle trough, the speed of the scraper chain, the particle size of the coal particles, and the laying inclination and the transportation efficiency of the scraper conveyor and the stress and deformation characteristics of the middle trough during transportation [8]. Yang applied EDEM to simulate the coal bulk conveying process in the scraper conveyor. The study found that the pressure in the medium plate when the coal bulk is conveyed at full load is about twice the weight of the coal being conveyed [9]. Zhou et al. used DEM to simulate and analyze the effect of particle shape and swirl strength on the coal fragmentation in pneumatic pipeline transportation [10]. Chen et al. used EDEM to simulate the process of conveying coal bulk by the scraper conveyor, and obtained the influence of chain speed, laying angle, hardness and particle size on the wear of the middle trough [11].

Wang et al. used DEM to simulate the coal conveying process of scraper conveyor, and studied the influence of scraper chain speed, coefficient of static friction between the particles and the coal spreading angle on mass flow, and the influence of scraper chain speed, static friction coefficient and scraper inclination angle on mass flow. The relationship among conveyor efficiency, chain speed, static friction coefficient and chute wear was studied by the method of combining DEM with FEM. Also, the stress and deformation characteristics of chute in the process of transportation were studied [12, 13]. Li et al. used EDEM to simulate the movement state of coal bulk conveyed by scraper conveyor. They found that when the transport state was stable, the average speed of coal particles accumulated under heavy accumulation condition was lower than that under light accumulation condition, and when the local plate was inclined, the transport efficiency of scraper conveyor would be reduced [14]. Li et al. combined the DEM with the FEM, and designed the middle trough according to the bionic principle to optimize the wear condition of the middle trough. The results show that the pit surface is easy to deform, resulting in more compression and bending strain energy. The pit has the function of releasing stress, thus improving the wear resistance of the non-smooth surface [15]. Xia et al. combined the DEM with the wear model, studied the wear mechanism and wear law of the central groove by observing the change of coal particle position during the wear process, and predicted the central groove of the scraper conveyor under various working conditions and bulk wear under coal characteristics [16, 17].

2.1.2 Mining Machinery

For mining machinery, DEM is widely used in the fields of mechanical structure, process improvement and working condition optimization. The biggest advantage of DEM in this aspect is the analysis of particle motion state and structure stress, which has a greater role in promoting the improvement of coal mining machinery and mining technology. Huang et al. simulated the process of inserting bulk in the loader bucket in EDEM, and the results showed that when the bucket inserted bulk of different particle sizes, the formation of the compaction belt was not much related to the particle size, but the larger the particle size, the less obvious the sliding cross-section of the particles [18]. Curry et al. used the collaborative simulation between DEM and multi-body dynamics to better simulate the complex motion and load transfer process between equipment and materials [19]. Zhao et al. used EDEM to simulate the performance of the coal bulk movement, coal loading rate and load in the prototype roller and model spiral roller during cutting, and proved the reliability of the spiral roller [20].

In view of the coal-breaking process, Jia et al. used PFC^{3D} to simulate the cutting process of the pick-shaped cutter, and analyzed the connection between the coal-cutting thickness and the cutting force. The results showed that the average cut force increased with the increase of cut thickness, the load fluctuation struck the tendency of first decrease and then increased, and obtained the cut performance in 15–20 mm cut thickness when relatively good [21].

In view of the different coal-releasing processes, Wang et al. used the principles and methods of PFC^{2D} to study the phenomenon of natural collapse, and the research showed that the PFC^{2D} model could effectively simulate the mining process of natural collapse method, analyze its internal mechanical mechanism and visually reflect the specific morphological changes of the ore parameters in the process of collapse [22]. Wang et al. took the bottomless column segment collapse method as the research object, and used DEM to carry out a study on the influence of the change of the corner of the fan-shaped gun hole on the release process, the internal complex sports field and internal force field [23]. Hu simulated the process of coal caving on working face by PFC^{2D}. Based on the study of the effect of different coal caving steps, it was found that the best way of coal caving on the 2200 working face of Cheng Zhuang coal mine was to carry out twice mining and once caving at a step of 1.6 m [24].

For the filling particle flow, Li et al. established the particle flow model of the falling zircon filling body by using the PFC, simulated the whole dynamic development process of coal mining, roof collapse, filling and filling material compaction by using the test, and studied the influence of the filling rate, the influence of the macroscopic properties such as the bonding strength and elastic modulus of the filling material on the compaction characteristics of the filling body [25].

In view of the top coal recovery ratio under different conditions, Xie et al. installed a vibration device on the protective beam of the top coal caving release support, so that the arch structure formed by falling top coal caving under the action of gravity was effectively disintegrated, which improved the top coal recovery ratio, and they used the DEM to carry out numerical simulation, and verified the destructive effect of the vibration on the top coal release structure formed in the process of the arch structure [26]. Li combined with the engineering practice of fully mechanized caving mining in the steep and extra thick seam of Aknifehe mine, studied the influence of different mining and caving ratio and different caving step distance on the top coal recovery ratio, and optimized the mining technology [27].

2.1.3 Other Coal Machine Equipment

With the continuous development and improvement of DEM, DEM is gradually widely used in various fields. In the coal machine equipment, the DEM is widely used in transportation equipment and mining machinery as well as other coal machine equipment.

As an important equipment in the field of coal, the coal silo whose storage and discharge performance has a decisive impact on the coal production capacity of major coal mines is an indispensable part of coal mining area. Taking the coal silo and coal bulk as the research object, the researchers use DEM to analyze the flow state and mechanical behavior of the coal bulk in the bunker and the influence of them on the silo, and put forward scientific suggestions for the application of the coal silo.

Yu et al. used DEM to simulate the discharge process of the silo, and by numerically simulating the internal pressure of the silo and the change of storage flow they analyzed the mechanical behavior and motion state involved in the loading and unloading process, analyzed the effect of particle density and material density on the results, and verified the reliability of the simulation results by experiment [28]. Yu et al. carried out a deeper discussion on this basis, and by comparing the results of simulation and physics experiments, it was proved that it was feasible to simulate and analyze the pressure of silo wall and the flow of storage materials by using the DEM [29]. Masson and Martinez used DEM to simulate the effect of the different mechanical properties of particles on particle flow and pressure in the flat rectangular silos, and found that the main factors were the friction and contact stiffness between particles [30]. Cleary and Sawley used DEM to simulate the discharge process of cylindrical silos with different discharge ports, deeply explored the effect of particle shape on the discharge state under three-dimensional conditions and proved the accuracy of simulating the particle flow process with DEM [31]. Xiao et al. analyzed the arching phenomenon in coal silo blocking accidents from the aspects of coal adhesion mechanism, coal arch force and coal silo structure, and pointed out the advantages and disadvantages of various methods by summarizing the current commonly used arch-breaking methods [32]. Wang et al. used the DEM to simulate the discharge flow in the discharge process, the dynamic force of the silo walls and the internal pressure distribution of the material to reveal the internal mechanism of the macro deformation and cracks in the coal silo wall, and found that the peak pressure was weakened and the arch structure was destroyed [33]. Zhang et al. used the Jenike shear tester and DEM to study the material characteristics of different material silo walls and the effect of different coal particle sizes on the friction angle of the coal powder moving wall [34]. Chen et al. simulated the unloading process of coal storage in the silo with the particle discrete element software, and then explored the internal mechanism of particle flow in the silo, and used the method of marking the particles to track the speed and displacement change of individual particles in the flow process, and obtained the static/dynamic side pressure of the silo wall and the dynamic pressure change of the hopper on a plane [35]. Gonzalez-Montellano et al. simulated the filling and unloading behavior of storage material silos and hoppers by two three-dimensional discrete element models, and the results reflected the potential of DEM model in studying the behavior of materials in silos and hoppers [36]. Zhen et al. used PFC^{3D} to simulate the flow state and the pressure distribution of the coal bulk in the discharge process of the coal silo and the internal pressure distribution of the coal bulk, and the results showed that the horizontal side pressure of the silo wall was significantly greater than static in dynamic, and the dynamic pressure fluctuated up and down with the outflow of particlcs [37]. Wang et al. used EDEM to establish an analytical model for a certain type of coal silo and coal bulk, simulated the bulk flow characteristics of coal silo during unloading, and analyzed the influence of coal moisture content, wall friction coefficient, the inclination and caliber of the discharge hopper on the coal flow state and discharge speed [38].

2.2 Application of DEM in Agricultural Machinery Equipment Ripper

As a discontinuous numerical simulation method for the analysis of discrete matter, compared with the traditional finite element analysis, the DEM no longer regards soil and crops as continuum, but as discrete matter for analysis. Compared with continuum theory, DEM is more practical. The interaction between agricultural machinery and soil/crops is studied from the particle micro level by using the DEM, which makes the optimization and improvement of agricultural machinery a big step forward.

2.2.1 Loose Machinery

Many scholars studied the loosening machinery in agricultural machinery by DEM. Momozu et al. simulated the loosening process of soil under the action of shearing tools by using a discrete element model improved after adding viscosity to soil particles, which better reflected the soil loosening process in the real experiment [39]. Zhang and Li established a discrete element mechanical model of viscous soil to simulate the dynamic behavior of bonded soil under external forces. The results showed that the fragmentation and separation of bonded soil blocks during excavation simulation is basically the same as the actual viscous soil under external force, but the suitability of the model is two-dimensional model [40]. Van der Linde applied DEM to explore the farming resistance and operational effect of vibration-type subsoilers at different vibration frequencies, and found reasons for vibration-reducing resistance from the particle level [41]. Sun et al. used DEM to simulate the process of cone-through into the soil, and compared the simulation experiment with indoor test results and found that the penetration curve of particle agglomeration obtained by simulation is consistent with the results measured by laboratory test. Both results showed that the penetration resistance gradually decreases with the porosity increasing [42]. Li et al. used DEM to simulate the whole process of cone-through sand, discussed the movement trajectory of the specified particles at the starting position of the bed, analyzed the displacement field of the bed and finally obtained the distribution law of the velocity field and force chain of the bed at different depths of penetration [43]. Chen et al. established a soil model using the DEM, and simulated the soil adhesion effect using the particle–particle combination method. By comparing the simulation results of subsoilers in three different soil conditions with the field verification test results, it was found that in most cases, the correlation error was not more than 10% [44]. Tamas et al. established the discrete element model of soil particles with the data of direct cut-ting test and triaxial compression test as input parameters, simulated the working process of the subsoiler and compared it with the actual test. The results showed that the subsoiler achieved better results at the speed of 0.5–2.4 m/s and the entry angle of 30° [45]. Wang used DEM to study a rear-sweeping arc subsoiler with the simulation analysis of the subsoiler, and found that the rear-sweeping angle of 50° of subsoiler by the smallest operating resistance resulted in the best ground-breaking effect [46]. Li et al. modified the microscopic parameters of soil particles by means of three-axis compression test and created a discrete element model of the interaction between the subsoiler and the soil, and the working resistance curve of the three different depths of the subsoiler obtained by the simulation had a better match with the field experiment, indicating that it was feasible to study the working process of the subsoiler by using DEM [47]. Deng et al. used EDEM to study the whole operation process and farming resistance of subsoiler, analyzed the operation effect of subsoiler in different farming depth, working speed and soiling angle, compared the simulation analysis results with the field test results and found that the greater the depth of cultivation and work speed, the smaller the depth of the soil, and the greater the resistance of the soil [48]. Li et al. used DEM to simulate the subsoiling process of different subsoilers under different operating speed and working depth, and through comparative analysis, it was found that the simulation results were basically consistent with the field test results, and the study pointed out that the subsoiler with a rear skimming angle of 50° in the subsoiling process was the best performance, and the loose soil can be achieved without turning the soil [49].

2.2.2 *Seeding Machinery*

Many scholars have applied the DEM to the seed metering device and furrow opener of agricultural sowing machines.

In the aspect of seed metering device, Li established a two-dimensional DEM model of the inside-filling seed metering and wheat seeds, simulated the working process of the seeder and analyzed its performance through the two-dimensional DEM software independently developed [50]. Shao carried out simulation and experimental research on the particle movement of the population during the working process of the permanent magnet magnetic roller precision seed metering device through EDEM [51]. Yuan et al. designed a spoon wheel seed metering device, and simulated the process of soybean seed metering by using discrete element software [52]. They designed a precision seed metering device of the hole wheel, and analyzed the influence of the chamfer of type hole, the installation angle of the seed cleaning brush and the rotation speed of the seed wheel on the seed metering performance with EDEM [53]. Xu et al. carried out the orthogonal virtual seed metering experiment on the seed metering performance of the seed metering device by using orthogonal experimental design and EDEM, and carried out regression analysis of the test results, thus optimizing the optimum seed metering performance parameters of the inclined disc spoon soybean seed metering device [54]. Hou et al. put forward a claw-type circulating single seed collecting device based on the design idea of taking more and leaving one to solve the problem of low single seed rate in mechanized planting of garlic [55]. Huo measured and calibrated the physical parameters and contact parameters of agaricus bisporus seeds, and optimized the structural parameters of seed metering device by simulation test with EDEM [56].

In terms of the trench opener, Yu et al. established a two-dimensional discrete element analysis model of core share furrow opener and soil particles with two-dimensional CAD. The working resistance of furrow opener under different conditions was studied based on DEM. The resistance of the two furrow openers was compared by changing the working speed, soil moisture content, furrow depth and other parameters, and they carried out soil bin test. The results showed that the relative error between simulation and real test is 10–20% [57]. Pan used DEM to explore the interaction between the core share opener and the soil, and carried out the structural optimization design of the opener. The results showed that the optimal structural parameters of the opener are as follows: the entry angle is 15°, the bevel angle is 55° and the width is 50 mm [58]. Yang established a trench opener-soil dynamics model based on DEM, and explored the effects of the key structural parameters of the opener on the performance of the opener and soil disturbance. The optimal structural parameters are as follows: the entry angle is 50°, the entry gap angle is 7° and the bevel angle is 50°. The experimental results showed that the optimized trenching resistance is reduced from 5.80 to 6.14%, and the soil disturbance is also reduced [59]. Yang et al. used uniform test and DEM to determine the optimal combination of the key structural parameters of the opener, such as soil entry angle, soil clearance angle and bevel angle to reduce the running resistance and disturbance to the soil of the trench opener of the pennisetum giganteum planter. The results showed that the optimal trench angle, clearance angle and bevel angle are 45°, 3° and 60°, respectively [60]. Aiming at the problem of soil backflow of ditcher, He et al. designed a diversion device and simulated it with DEM [61].

2.2.3 Harvesting Machinery

Many scholars have studied the application of DEM in agricultural harvesting machinery. Yuan compiled a two-dimensional vibration screening simulation program, and studied the effects of vibration frequency, screen surface inclination angle, rear suspender length and screen surface area on soybean particle screening efficiency [62]. Jiang et al. conducted a coupling study on the movement of the mixture of rice particles and short stalks in the inertial cleaning chamber of the combine harvester before cutting. The effects of inlet air velocity on grain distribution and velocity (length and height direction) and particle density on material trajectory were analyzed. The results showed that on increasing the inlet air velocity we can improve the cleaning performance of the device, but it will aggravate the turbulence at the back end of the cleaning room, and then increases energy consumption [63]. Yu simulated the screening process of the plot grain combined (plane reciprocating and fisheye screen). The effects of vibration parameters (frequency, amplitude and vibration direction) and structural parameters (sieving hole, screen surface inclination, sieve size and spacing) on the screening performance (screening efficiency, cleaning rate and loss rate) of wheat grains and short stalks were studied [64]. Lu simulated the separation process of wheat grain and short stalk in plane reciprocating vibration screening device, studied the influence of wind speed on material displacement distribution and longitudinal average velocity, obtained the influence of amplitude, frequency, fan inclination angle and wind speed on cleaning efficiency through orthogonal experiment, and obtained the best parameter combination of cleaning efficiency [65]. Liu et al. established the soil-carrot model of the extraction process according to the discrete element theory, analyzed the disturbance effect of the extraction on the soil and discussed the influence of the main working parameters of mechanical extraction on the mechanical drawing force of carrots. The results showed that the maximum pullout force appears at the beginning of the process, the disturbance area is inverted conical and the pullout force was the smallest when the pullout force was along the direction of carrot growth, which provides a theoretical reference for the development of root vegetable harvesting machinery [66].

Aiming at the problem of high injury rate of sweet potato harvester, Gao et al. made a single factor simulation analysis of its mechanism by EDEM, and analyzed the simulation results with the parameters of normal stress and transmission efficiency of sweet potato as the index parameters. The results showed that the operation effect of two-stage delivery and separation mechanism was significantly better than the original delivery and separation mechanism. The average loss rate of the one-stage operation mechanism is as high as 73.9%, and the average loss rate of the two-stage operation mechanism is 11.0%, which proves the feasibility of two-stage transport and separation mechanism and the innovation of TRIZ theory [67]. Wen et al. calibrated the physical parameters and contact parameters of cane, sugarcane tail and sugarcane leaf, and then analyzed the mechanism of removing impurities of sugarcane harvester with the method of CFD-DEM coupling, which provided a theoretical basis for optimizing the structure parameters of the device [68].

2.2.4 Other Agricultural Machinery

DEM is also often used in other agricultural machinery, such as fertilizing machine, returning machine, kneading machine, building machine and so on.

The fertilizer dispenser is a key component of the fertilizing machine, and scholars have done a lot of research in this aspect. In order to improve the fertilizer discharge performance of the fertilizer applicator in the greenhouse, Sun et al. used EDEM to carry out the orthogonal virtual experiment on the fertilizer discharge performance. Taking the consistency variation coefficient of fertilizer discharge of each row, the stability variation coefficient of total fertilizer discharge and the fertilization breaking rate as the evaluation index of fertilizer discharge performance, the optimal parameter combination was obtained: when the advance speed was 7.35 km/h, the rotational speed was 48.76 rpm and the outer diameter of the grooved wheel was 53.28 mm, the evaluation indicators were, respectively, 1.08, 0.93 and 0.14% [69]. Wu et al. designed a duck tongue deep fertilizer applicator, established a discrete element simulation model of the deep fertilizer applicator, used EDEM to simulate the fertilizer discharge process under different operating parameters and discussed the effects of fertilizer application type, depth, advance speed and soil moisture content on fertilizer discharge and working resistance [70]. Xue et al. simulated and analyzed the fertilizer discharge process of spiral fertilizer discharge, found that the model structure of spiral fertilizer drainer was reasonable, the stability and uniformity of the drainage process were good and analyzed the influence of diameter of spiral blade, pitch and speed of the drainage shaft on the fertilizer discharge performance [71, 72]. Shen et al. established the discrete element model of outer groove wheel fertilizer discharger and fertilizer particles, took the fertilizer amount and the variation coefficient of fertilization evenness as the evaluation index of fertilization effect and simulated the fertilizer discharge process under four slopes by EDEM [73]. Yang et al. put forward the method of improving the outer groove wheel, and simulated the fertilizer discharge process of different outer groove wheel fertilizer discharge to improve the fertilization accuracy and stability of the two-variable fertilizing machine. The results showed that the improved outer groove wheel c has the best fertilizer discharge stability and the smallest coefficient of variation [74].

In the field of returning machines, Zhang designed an energy-saving straw returning machine, simulated the interaction among straw, soil and rotary knife roller by DEM, and carried out machine experiments in the field [75]. Wang optimized the knife roller assembly to solve the problem of high consumption and low efficiency of returning straw to field by reverse rotary tillage. The results showed that the knife performance is the best when the knife width is 79 mm, the bending line angle is 56° and the bending angle is 81° [76].

Based on the theories of DEM, MBD and CFD, Song used corn straw as processing material to simulate the process of corn straw crushing and filamentous material discharge, which provided some theoretical reference for visual dynamic research and equipment optimization design of 9FH-40 straw kneading machine [77]. Ma et al. used HSCM and LCM contact model to carry out central combination simulation experiments with soil accumulation angle as the index, soil particle recovery coefficient, static friction coefficient, rolling friction coefficient and soil adhesion energy density as factors. The research results can provide theoretical reference for the development of grapevine anti-cold soil cleaning machine suitable for sandy soil texture [78]. Ku et al. put forward a new tool arrangement method, and studied the influence of different tool arrangement on tool shaft working performance and soil crushing effect by EDEM to improve the working quality and efficiency of rotary ploughing tool shaft [79].

References

1. Qiu X, Kruse D (1997) Analysis of flow of ore materials in a conveyor transfer chute using discrete element method. In: Proceedings of the 1997 ASME-ASCE-SES Joint Summer Meeting, June 29, 1997–July 2, 1997. Evanston, IL, USA: ASCE, pp 395–404
2. Katterfeld A, Groger T (2007) Application of the discrete element method—part 4: Bucket elevators and scraper conveyors. Bulk Solids Handl 27(4):228–234
3. Yang Y (2011) Research on pressure distribution of handled bulk material and resistance calculation method. Dissertation. Northeastern University. (in Chinese)
4. Piao X, Guo Y (2012) Application of discrete element simulation technology to belt conveyor. TECHNO SCI COAL 40(03): 87–90. (in Chinese)
5. Ma M, Liu Y (2013) Simulation research of loads on scraper of road header. Coal Mine Machinery 11:71–73 (in Chinese)
6. Zhou W, Wei H (2013) Simulation and analysis of belt conveyor based on edem software. Coal Mine Machinery 34(5):89–91 (in Chinese)
7. Zidek M, Zegzulka J, Jezerska L, Hlosta J, Necas J (2015) Optimisation of geometry of the chain conveyor carriers by DEM method. Inz. Miner. 2:143–148
8. Cai L, Wang X, Li B, Yang Z (2016) Analysis on conveying efficiency and stress and deformation of middle pan during conveying process. Machinery Design & Manufacture 12:172–176 (in Chinese)
9. Yang M (2017) Dynamic characteristics research on internal bulk boal load in middle pan of scraper conveyers. Dissertation. Taiyuan University of Technology
10. Zhou J, Liu Y, Du C, Liu S (2017) Effect of the particle shape and swirling intensity on the breakage of lump coal particle in pneumatic conveying. Powder Technol 317:438–448
11. Chen Z, Wang X, Wang S, Wang S, Yang Z (2017) Wear analysis on middle pan of scraper conveyor. Coal Technol, (2). (in Chinese)
12. Wang X, Li B, Wang S, Yang Z, Cai L (2017) The transporting efficiency and mechanical behavior analysis of scraper conveyor. Proc Inst Mech Eng Part C-J Eng Mech Eng Sci 232(18): 3315–3324
13. Wang X, Li B, Yang Z (2018) Analysis of the bulk coal transport state of a scraper conveyor using the discrete element method. Strojniski Vestn J Mech Eng 64(1): 37–46

14. Li B, Wang X, Chen Z, Yang Z (2018) Research on simulation of bulk coal conveying state. TECHNO COAL, 37(02): 251–253. (in Chinese)
15. Li, B., Wang, X., Xia, R., Yang, Z., Liu, Y., & Li, T. (2018). Research on the bionic design of the middle trough of a scraper conveyor based on the finite element method. Proc. Inst. Mech. Eng. Part C-J. Eng. Mech. Eng. Sci., 233(9), 3286–3301
16. Xia, R., Wang, X., Li, B., Wei, X., & Yang, Z. (2019). The prediction of wear on a scraper conveyor chute affected by different factors based on the discrete element method. *Proc.* Inst. Mech. Eng. Part C-J. Eng. Mech. Eng. Sci., 233(17), 6229–6239
17. Xia R, Wang X, Li B, Wei X, Yang Z (2019) Discrete Element Method- (DEM-) Based Study on the Wear Mechanism and Wear Regularity in Scraper Conveyor Chutes. Math. Probl. Eng. 2019:1–12
18. Huang PP, Xiao XZ (2013) Simulation study the movement of materials in loader shovel working process based on edem. J. Adv. Mater. Res. 655–657:320–325
19. Curry, D.R., &Deng, Y. (2016). Optimizing Heavy Equipment for Handling Bulk Materials with Adams-EDEM Co-simulation. C. International Conference on Discrete Element Methods. Springer, Singapore
20. Zhao L, Fan J, Wang, Y. L (2019) Exploring Dynamic Cutting Process of Shearer's Drum. J. MACH SCI TECHNOL. 38(03):64–69 (in Chinese)
21. Jia, J., Wang, Y., Yang, Z., & Xu, Chan. (2017). Analysis of coal breaking force under different cutting thickness of pick. *J. Coal Engineering*, 49(9):122–125,129
22. Wang L, Gao Q, Wang J, Fang Z (2007) Numerical simulation of natural caving method based on particle flow code in two dimensions. J. Journal of Engineering Sciences 29(6):557–561 (in Chinese)
23. Wang P, Yang T, Liu X (2010) Particle Flow Numeral Simulation Investigationon Influence of Lateral Opening Angleon Ore Drawing with Sublevel Pillarless Caving. J. Metal Mine 3:12–16 (in Chinese)
24. Hu Y (2017) Research on coal caving step of fully mechanized caving step distance at fully mechanized caving face based on PFC2D. J. China Coal 43(3):70–73 (in Chinese)
25. Li, X., Xu, J., Zhu, W., Zhuang, D. (2008). Simulation of back fill compaction character by particle flow code. *J. J CHINA COAL SOC.*,33(4), 373–377. (in Chinese)
26. Xie, Y., & Zhao, Y. (2009). Numerical simulation of the top coal caving process using the discrete element meth-od. *J. Int. J. Rock Mech. Min. Sci.*, 46(6), 983–991
27. Li F (2010) Simulation analysis of horizontally layered fully mechanized top coal caving effect in steep and extra thick coal seam. J. Coal Engineering. 03:85–87
28. Yu L, Xing J, Tu J, Yang J (1999) Numerical Simulation of Internal Forces and Flow Fields in Silos and Experimental Verification. J. Journal of Yantai University: Natural Science and Engineering Edition 04:23–30 (in Chinese)
29. Yu L, Xing J (2000) Discrete Element Method Simulation of Forces and Flow Fields During Filling and Discharging Materials in Silos. J. Journal of Agricultural Engineering 16(4):15–19 (in Chinese)
30. Masson S, Martinez J (2000) Effect of particle mechanical properties on silo flow and stresses from distinct element simulations. J. Powder Technol. 109(1–3):164–178
31. Cleary PW, Sawley ML (2002) Dem modelling of industrial granular flows: 3D case studies and the effect of particle shape on hopper discharge. J. Appl. Math. Model. 26(2):89–111
32. Xiao L, Li Z, Xiao H (2007) Analysis of Causes and Solutions of Arch-forming in Coal Bunker. J. Coal Mine Electromechanics 5:16–18 (in Chinese)
33. Wang Y, Guo M, Wu L (2008) Distinct Element Numerical Simulation of Dynamic Force Field in Raw Coal Silo. J. Modern Electric Power. 06:69–72 (in Chinese)
34. Zhang, Y., Shen, X., Tang, X., & Yuan, Z. (2010). Wall friction characteristics of pulverized coal by experiment and numerical simulation. *J.J SOUTHEAST U: Nat Sci Ed*, (01), 137–142. (in Chinese)
35. Chen X, Xu Z, Qin G, Zhang C (2011) Distinct element method simulation of discharging in coal silos. J. China Powder Technology 17(005):61–64 (in Chinese)

36. González-Montellano C, Gallego E, Ramírez-Gómez A, Ayuga F (2012) Three-dimensional discrete element models for simulating the filling and emptying of silos: analysis of numerical results. J. Comput. Chem. Eng. 40:22–32
37. Zhen H, Wang X, Yang Z (2013) Distinct element simulation on lateral pressure during coal discharging process of coal silo based on PFC^{3D}. J. Coal Engineering. S1:123–125 (in Chinese)
38. Wang, X., QIN, Y., TIAN, Y., YANG, Xin, & Yang Z. (2015). Analysis on flow features of bulk coal during coal unloading period based on EDEM. *J. COAL SCI TECHNO*, (05), 133–137. (in Chinese)
39. Momozu, M., Oida, A., Yamazaki, M., & Koolen, A.J. (2002). Simulation of a soil loosening process by means of the modified distinct element method. *J.J. Terramech.*, 39(4), 207–220
40. Zhang, R, & Li, J. (2006). Simulation on mechanical behavior of cohesive soil by distinct element method. *J.J. Terramech.*, 43(3), 303–316
41. Van der Linde, J. (2007). Discrete element modeling of a vibratory subsoiler. Dissertation. University of Stellenbosch
42. Sun P, Gao F, Li W, Yao S, Gong Y (2009) Discrete Element Simulation of Cone Penetration Process. J. Journal of Agricultural Machinery 40(1):184–188 (in Chinese)
43. Li Y, Wu T, Lin J, Xu Y (2012) Influence of penetrating cone on motion characteristics of sandy soil particle using discrete element method. J. Journal of Agricultural Engineering 024:55–61 (in Chinese)
44. Chen Y, Munkholm LJ, Nyord T (2013) A discrete element model for soil–sweep interaction in three different soils. J. Soil Tillage Res. 126:34–41
45. Kornél T, István JJ, Mouazen AM (2013) Modelling soil-sweep interaction with discrete element method. J. Soil Tillage Res. 134(8):223–231
46. Wang, Y. (2014). Simulation Analysis of Structure and Effect of the Subsoiler Based on DEM. Dissertation. Jilin Agricultural University
47. Li B, Liu F, Mu J, Chen J, Han W (2014) Distinct element method analysis and field experiment of soil resistance applied on the subsoiler. J. Int. J. Agric. Biol. Eng. 7(1):54–59
48. Deng J, Hu J, Li Q, Li H, Yu T (2016) Simulation and experimental study on the subsoiler based on EDEM discrete element method. J. China Agricultural Machinery Chemistry News 37(04):14–18 (in Chinese)
49. Li L, Li G, Zhang Y, Yuan H (2018) Simulation analysis of working process and effect of circular arc subsoiling shovel by discrete element method. J. Jiangsu Agricultural Sciences 46(13):201–204 (in Chinese)
50. Li Z, Yu J, Zhang W, Fu H (2011) Simulation analysis of working process and performance of inside-filling seed metering device by discrete element method. Transactions of the Chinese Society of Agricultural Engineering 27(11):32–36 (in Chinese)
51. Shao, X. (2011). Particle kinematic simulation and experimental study on the process of magnetic precision seed-meter sucking seeds based on DEM. Dissertation. Jiangsu University
52. Yuan G, Zhang T, Liu Y (2017) Design of a spoon wheel type seed metering device and simulation of soybean seeds by discrete element method. Journal of Agricultural Mechanization Research 39(11):25–29
53. Yu J, Lu C, Wei R, Fu L, Li K, Tang Z, Wang F (2018) Simulation test of performance of wheat precision seed-metering device based on discrete element method. Jiangsu Agricultural Sciences 46(08):225–228 (in Chinese)
54. Xu J, Cai Z, Zhang G, Shen P (2019) Study on optimization of seed spoon for declined disc-scoop-type soybean seed metering device based on the discrete element. Jiangsu Agricultural Sciences 47(04):192–196 (in Chinese)
55. Hou J, Wang H, Niu Z, Xi R, Li T (2019) Discrete element simulation and experiment of picking and clearing performance of garlic seed-picking device. Transactions of the Chinese Society of Agricultural Engineering 35(24):48–57 (in Chinese)
56. Huo, X. (2019). The experimental research on discrete element analysis parameters of the agaricus bisporus seeds and the seed metering device optimization. Dissertation. Chinese Academy of Agricultural Sciences

57. Yu J, Qian L, Yu W, Pan S, Fang Y, Fu H (2009) DEM analysis of the resistances applied on furrow openers. Transactions of the Chinese Society for Agricultural Machinery 40(06):53–57 (in Chinese)
58. Pan, S. (2015). Research on the optimization design and the experiment of the core ploughshare furrow opener based on the discrete element method. Dissertation. Jilin University
59. Yang, W. (2019). Structure optimization and experimental study of the pennisetum.sp opener based on the discrete element method. Dissertation. Fujian Agricultural and Forestry University
60. Yang W, Zheng S, Liu B (2019) Stimulation and optimization of Juncao opener structural parameters using discrete element method. Journal of Fujian Agriculture and Forestry University (Natural Science Edition) 48(05):673–680 (in Chinese)
61. He, B., Qi, J., Meng, H., Lin, Y., Zhou, X., & Zhao, W. (2019). Performance analysis and parameter optimization of diversion device of ditcher based on EDEM. *Journal of Agricultural Chinese Mechanization*, 40(10), 25–29 + 41. (in Chinese)
62. Yuan, L. (2011). Simulation analysis of screening process of ZL-2 model combine harvester based on 2D DEM. Dissertation. Jilin University
63. Jiang E, Sun Z, Pan Z, Wang L (2014) Simulation and experiment of grain movement in the separation chamber of harvester based on CFD-DEM. Transactions of the Chinese Society for Agricultural Machinery 45(04):117–122 (in Chinese)
64. Yu, B. (2016). Screening performance research and optimization of the plot combine harvester based on EDEM. Dissertation. Xihua University
65. Lu, X. (2017). Simulation analysis and optimization of wheat harvester cleaning device based on CFD - DEM coupling. Dissertation. Shandong Agricultural University
66. Liu Q, Xu L, Liu X, Lv Z, Song Y, Li C (2018) Research on dynamic behavior during mechanic pulling of carrot based on distinct element method. Journal of Chinese Agricultural Mechanization 39(06):61–65 (in Chinese)
67. Gao G, Li B, Yang D, Xie H (2019) Optimization of potato soil separation mechanism based on discrete element method and TRIZ theory. Modern Agricultural Equipment 40(05):10–18 (in Chinese)
68. Wen X, Yang W, Guo W, Zeng B (2020) Parameter determination and validation of discrete element model of segmented sugarcane harveste for impurity removal. Journal of Chinese Agricultural Mechanization 41(01):12–18 (in Chinese)
69. Sun, C., Cao, L., Shi, Z., Chen, X., & Huang, S. (2019). Analysis on fertilizer performance of fertilizer machine based on discrete element method. *Journal of Chinese Agricultural Mechanization*, 40(03), 44–50 + 127. (in Chinese)
70. Wu X, Wang F, Zhang F, Zhang Z, Li X (2020) Analyzing on factors influencing fertilizer discharge and resistance of deep fertilizing device base on EDEM. Journal of Agricultural Mechanization Research 42(03):175–181 (in Chinese)
71. Xue Z, Zhao L, Wang F, Wang S, Song G, Wang G (2020) Study on the performance of spiral fertilizer distributor based on EDEM. J Agricult Mechanizat Res 42(06):87–95 (in Chinese)
72. Xue Z, Zhao L, Wang F, Wang F, Wang G, Pan R (2019) Performance simulation test of the spiral fertilizer distributor based on discrete element method. J Hunan Agricultural University (Nat Sci) 45(05):548–553 (in Chinese)
73. Shen P, Yao Y, Zheng M, Kong H, Shen D, Wei G, Ning W (2019) Simulation and optimization of potato fertilizer dispenser in mountainous area based on discrete element method. Jiangsu Agricultural Sci 47(15):256–258 (in Chinese)
74. Yang H, Zhang L, Dong W, Fang Y, Hu D, Yang D (2019) Analysis and experiment of fertilizer device for bivariate fertilizer based on discrete element method. Machine Design Res 35(05):179–183 (in Chinese)
75. Zhang J (2018) Simulation optimization and experimental research of straw returning machine based on discrete element method. Dissertation. Anhui Agricultural University
76. Wang J (2019) Optimum design and experimental of rice straw deep–buried roller assembly based on discrete element method. Dissertation. Northeast Agricultural University
77. Song X (2018) Study on the operation process of the corn straw kneading machine based on discrete element method. Dissertation. Gansu Agricultural University

78. Ma S, Xu L, Yuan Q, Niu C, Zeng J, Chen C, Wang S, Yuan X (2020) Calibration of discrete element simulation parameters of grapevine antifreezing soil and its interaction with soil-cleaning components. Trans Chinese Soc Agric Eng 36(01):40–49 (in Chinese)
79. Ku H, Liu M, Cai H, Liu Y (2020) Performance analysis of rotary cutter of building machine based on discrete element method. J Chinese Agric Mechanizat 41(01):19–24 (in Chinese)

Chapter 3
Analysis of the Coal Bulk Transport State Based on DEM

3.1 EDEM Simulation Model Construction

3.1.1 Scraper Conveyor Model

In this paper, the scraper conveyor model SGZ880/800 was selected for research. The middle trough is the middle double chain. The chain specification is 34 × 126 (chain diameter × pitch). The sprocket is 7-tooth sprocket, and the driving mode is double sprocket, with the rated chain speed not exceeding 1.2 m/s.

In order to shorten the simulation time, reduce the requirements of computer performance and ensure the normal operation of simulation, the actual scraper conveyor structure was simplified. The head, tail, transition slot and baffle were removed, and only part of the structure of the lower middle trough of the horizontal section was reserved. The scraper guide slot was added to ensure that the state of coal bulk can be transported in the middle trough driven by the sprocket. The 3D geometric model of the middle trough was established by UG, and the simplified axial-side drawing of the three-dimensional geometric model of the middle trough is shown in Fig. 3.1.

X. Wang et al., *Engineering Applications of Discrete Element Method*, Engineering Applications of Computational Methods 4,
https://doi.org/10.1007/978-981-15-7977-6_3

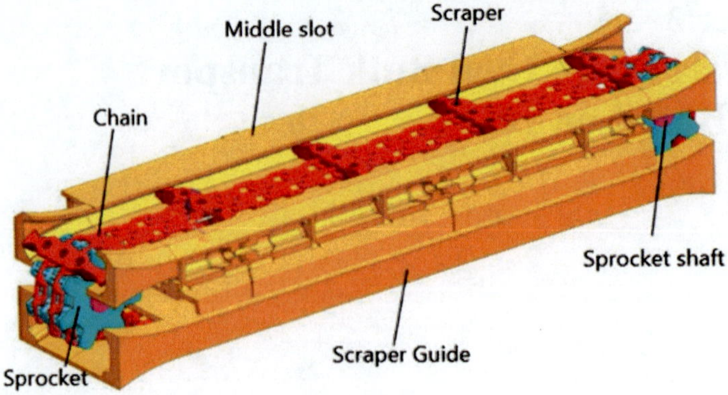

Fig. 3.1 Three-dimensional geometric model of the chute

3.1.2 Coal Bulk Model

In actual production, coal particles are irregular and have different shapes. In EDEM, many small balls can be stacked to simulate the shape of typical coal particles. The quality of particle model is directly related to the efficiency and accuracy of simulation. The closer the particle model is to the actual coal particle shape, the more accurate the simulation calculation. However, if too many small balls are used to accumulate coal particles, the particle production speed in the particle factory will be greatly slowed down. At the same time, the contact calculation between particles will become complex, and the simulation efficiency will be greatly reduced. According to the actual simulation requirements of the research, the shape, particle size and generation mode of coal particles used in various research contents are different. The specific situation of the coal particle model will be explained in each section.

The key parameters of DEM are the intrinsic parameters of coal bulk and geometry material and the contact parameters of contact model. These parameters are generally obtained by parameter calibration test. The parameters in the simulation come from the existing research results of the research team, and the specific parameters are shown in Tables 3.1 and 3.2 [1].

Table 3.1 The intrinsic parameters of coal and steel

Object	Shear modulus/Pa	Poisson's ratio	Density/kg/m^3
Coal	4.7×10^8	0.3	1229
Steel	8×10^{10}	0.3	7850

Table 3.2 Contact parameters of the contact model

Contact parameters	Value
Coefficient of restitution between coal and coal	0.64
Coefficient of restitution between coal and steel	0.65
Coefficient of static friction between coal and coal	0.329
Coefficient of static friction between coal and steel	0.46
Coefficient of rolling friction between coal and coal	0.036
Coefficient of rolling friction between coal and steel	0.032

3.2 EDEM Simulation Analysis of Transportation State Under Complex Conditions

3.2.1 Stable and Start-Stop Working Conditions

Firstly, the steady working condition is set as the reference state, and the relevant characteristics of the particle factory need to be set under the 'Factory' module in the EDEM during the simulation. In order to study the characteristics of particle transport state of coal bulk with different particle sizes, three particle factories with the same location and size were added to produce three particles with different particle sizes. The position and time of particle generation shall be exactly the same, so as to simulate the coal cutting with different particle size when the shearer was working. Three kinds of particles are set as 20, 35 and 60 mm spheres. And they are generated at the rates of 60, 80 and 80 kg/s, respectively.

3.2.1.1 Stable Working Condition

Particles were colored blue, green and red in the order of speed. Multiple particles were selected from three different particle sizes, and the motion track was shown in Fig. 3.2. In order to get the trend of speed change clearly, the coloring speed interval was set as 1–1.2 m/s.

When the particles are located at the bottom of the coal bulk, the closer the bottom of the middle trough, the greater the frequency of velocity change and the greater the velocity fluctuation. When the particles are in the upper part of the bulk, the velocity is relatively stable and the velocity fluctuation is small. But this change has no direct relationship with the particle size.

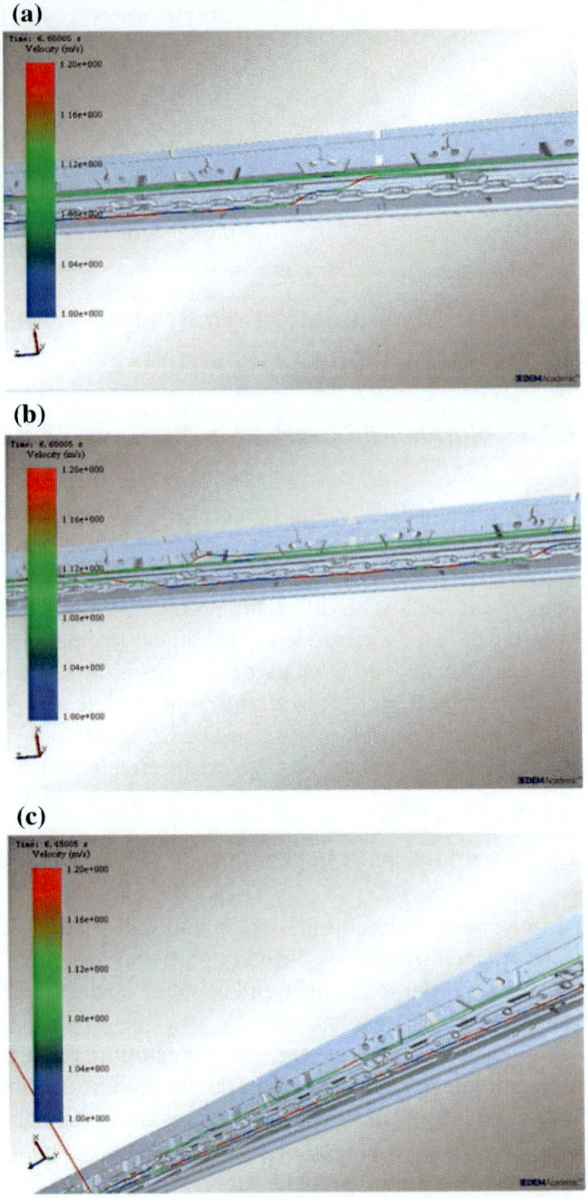

Fig. 3.2 Motion trail of the coal bulk

3.2.1.2 Stopping Condition

The dyeing function of the 'Post-processing' module of the EDEM was adopted to observe the movement state of particles before and after the sudden stop, so as to obtain the movement state of the coal bulk under the stopping condition. The

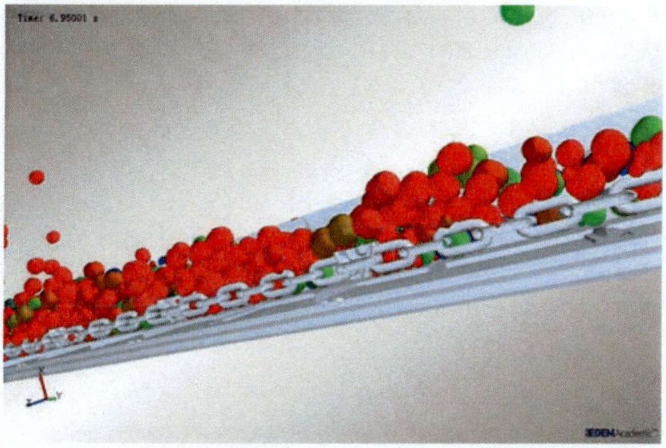

Fig. 3.3 Initial state of coal bulk in stopping condition

particles with different velocities are colored respectively, and the dyeing range is set as 0–1.1 m/s.

Figure 3.3 shows the profile of particle velocity when t = 6.95 s. It can be seen from the figure that the particle velocity of scraper conveyor before sudden stopping is almost 1.1 m/s that of scraper chain. The particle velocity profiles at t = 7.05 s, t = 7.15 s and t = 7.35 s are shown in Fig. 3.4.

It can be seen from the figure that after the scraper conveyor at t = 7 s stops suddenly, the speed of particles has changed greatly in a short time, and presents a layered speed change from down to up. The closer the bulk is to the bottom of the middle trough, the faster it falls to 0 m/s, and the slower the bulk is stationary away from the bottom of the middle trough.

3.2.1.3 Starting Condition

The starting condition was the same as the stopping condition, and the same research was on the movement of coal bulk in a very short time. In the post-processing module, the particles were dyed according to the particle speed by the dyeing system.

Figure 3.5 shows the profile of particle velocity at t = 4.85 s, which is the particle motion state diagram before the scraper conveyor suddenly starts. From the figure, the coal bulk in the scraper conveyor is in a static state. Figure 3.6 shows the particle velocity profile at t = 5.05 s, t = 5.15 s and t = 5.55 s.

It can be seen from the figure that after the scraper conveyor at t = 5 s is suddenly started, the particle speed also shows obvious layered speed change from down to up in a short time. The closer the coal bulk is to the bottom of the middle trough, the faster the speed increases. The farther away the coal bulk is from the bottom of the middle trough, the slower the speed increases. It can be seen from Fig. 3.6c that when t = 5.55 s, the movement state of coal bulk is basically the same as that under the previous stable condition, and the whole process is also very short. At the same time, it can be seen from Fig. 3.7 that the speed distribution of coal bulk after startup is also consistent with the speed distribution analyzed above.

Fig. 3.4 Change of coal bulk motion state after stopping

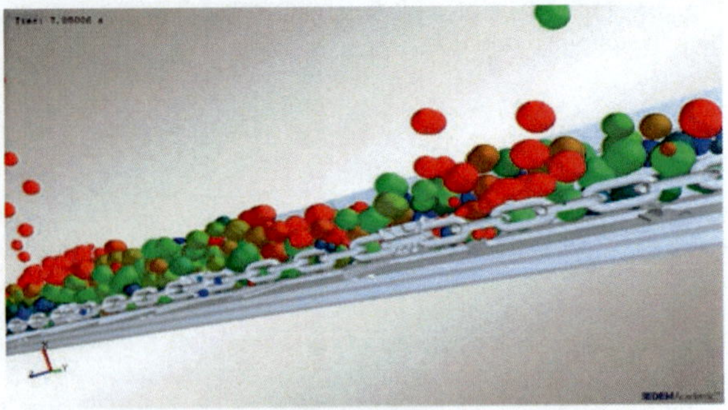

(a) At 7.05s

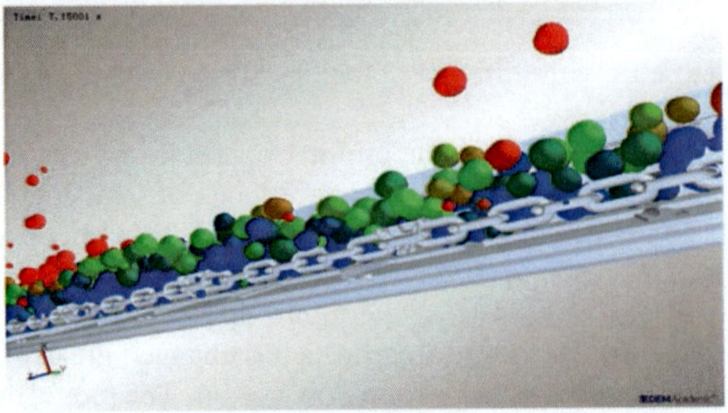

(b) At 7.15s

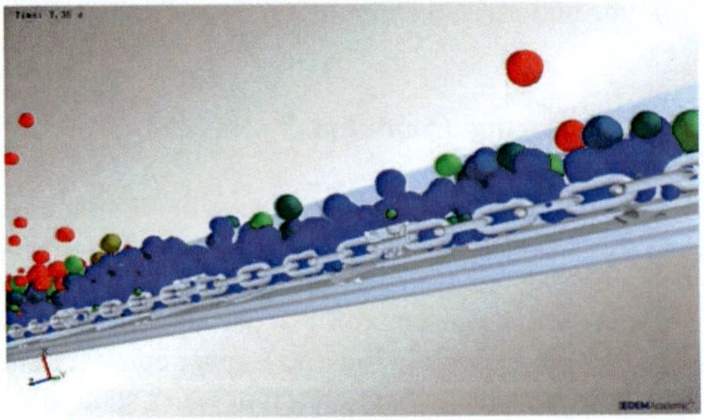

(c) At 7.35s

3.2.2 Material Accumulation Conditions

By using EDEM, the two working conditions of local light accumulation and local heavy accumulation of materials in the middle trough of scraper conveyor were simulated, and the movement behavior of coal bulk and the distribution pattern of coal bulk were analyzed under these two conditions. The specific implementation process was to set up a particle factory near the middle part of the middle trough, and

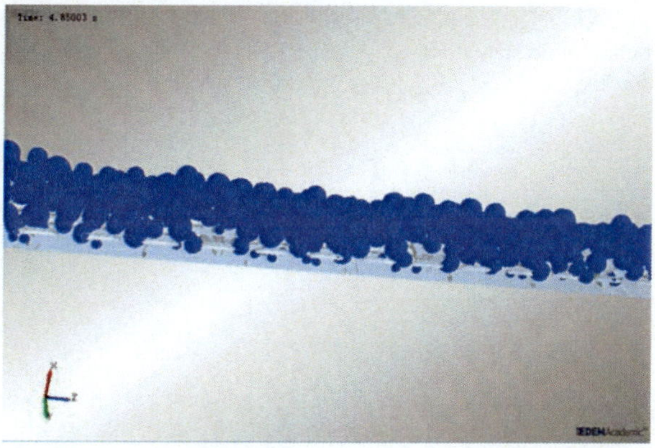

Fig. 3.5 Initial state of coal bulk in starting condition

simulated the material accumulation condition by generating particles in advance. When simulating light accumulation, 20 kg coal bulk were generated in advance, as shown in Fig. 3.8. When simulating heavy accumulation, 60 kg coal bulk were generated in advance. In the simulation study of accumulation conditions, the type of particles generating accumulation was particle of 35 mm. The laying angle of the middle trough was set as 0°, the wear constant and scraper chain speed were 1.5×10^{-12} m^2/N and 1.1 m/s, respectively, the traffic volume was set as 330 kg/s, and the simulation time was 10 s.

The distribution pattern of coal bulk at 2 s of light accumulation condition was shown in Fig. 3.9a, in which red represented the accumulation material and blue represented the normal transportation of coal bulk. Figure 3.9b shows the particle distribution at 7 s under slight accumulation condition. Through the comparison between 2 s and 7 s time, it could be seen that at the beginning of the movement, the coal bulk transported normally were distinguished from the accumulated particles obviously. As time went by, the accumulated red particles were dispersed by the normally transported blue particles. At this time, they were not mixed, and the red particles were distributed on both sides of the blue particles.

Figure 3.10 showed the particle distribution at 7 s under severe accumulation condition. It could be seen from the figure that the red particles were not completely dispersed by the blue particles at this time. It could be seen that the more serious the material was piled up, the longer the time for particles to be dispersed. In addition, it was observed from Fig. 3.10 that in case of heavy accumulation, the accumulated particles would be pushed on the ledge of middle trough along with the operation of the conveyor.

Fig. 3.6 Change of coal bulk motion state after starting

(a) At 5.05s

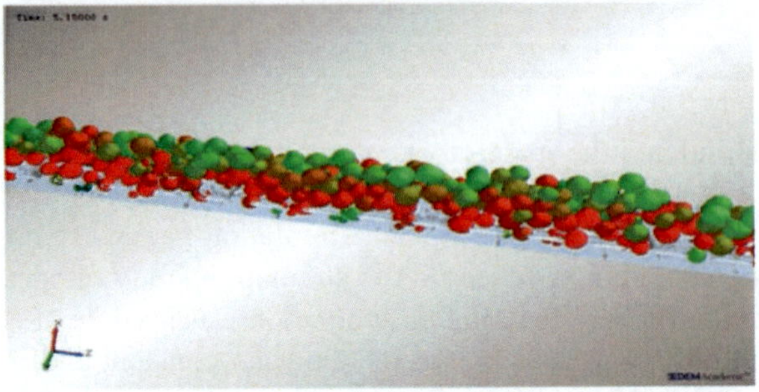

(b) At 5.15s

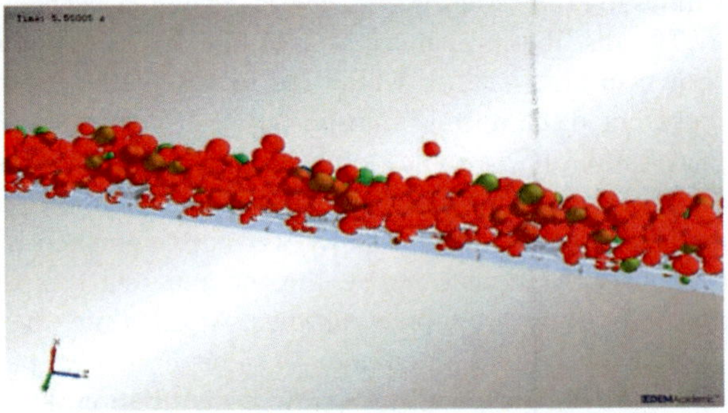

(c) At 5.55s

3.2.3 Gangue-Containing Working Conditions

In the actual fully mechanized working face, the coal is cut by the shearer and drops to the scraper conveyor for transportation, which must contain a certain amount of gangue, and its content is generally about 0–20%. In order to make the simulation closer to the actual production, this chapter analyzes the movement form and velocity distribution under the condition of gangue by establishing the coal bulk model with 7% gangue content. The scraper chain speed was set to 1.1 m/s, the laying angle was

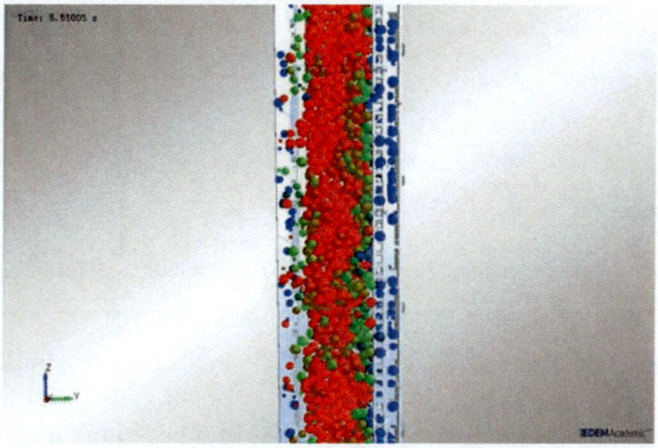

Fig. 3.7 Stable state of starting condition

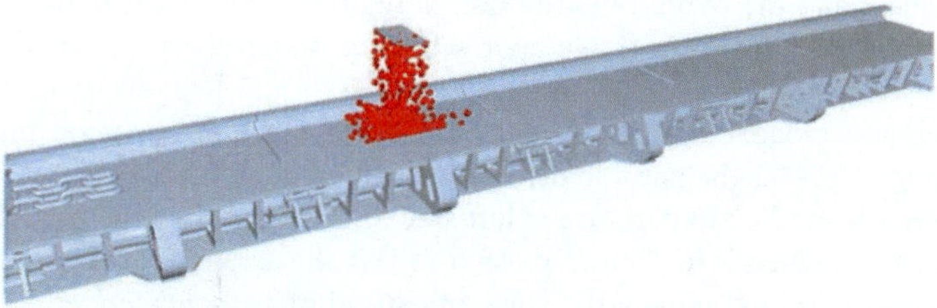

Fig. 3.8 Schematic diagram of material accumulation

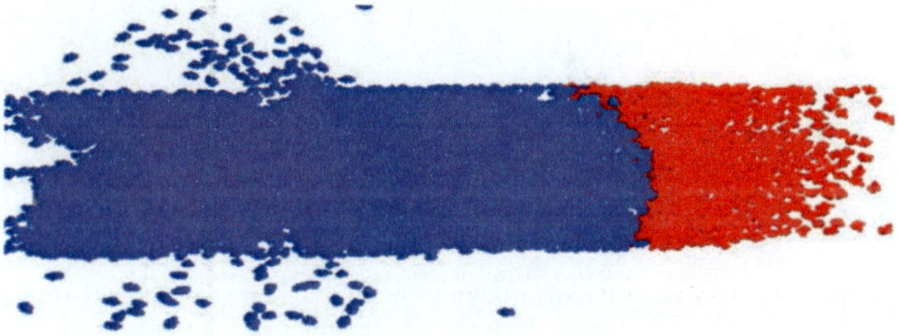

(a) At 2s

(b) At 7s

Fig. 3.9 Distribution of coal bulk under light accumulation condition

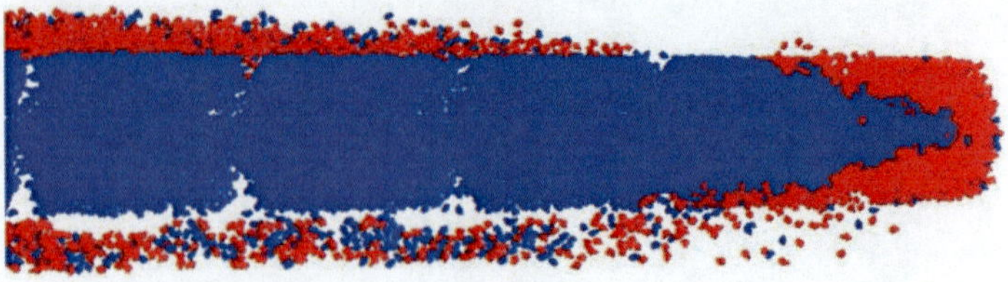

Fig. 3.10 Distribution of coal bulk at 7 s under severe accumulation condition

set to 0°, the wear constant was set to 1.5×10^{-12} m^2/N and the simulation time was set to 13 s.

Figure 3.11 shows longitudinal profile of the middle trough under the condition of gangue, in which light yellow represents coal bulk and red represents scraper and chain. It can be seen from the figure that the coal bulk forms a small slope pile every other distance, and the overall distribution is regular. The lowest part of the slope pile appears on the left side of the scraper, while the highest part appears on the right side of the scraper.

The velocity cloud diagram of coal bulk along the transportation direction is shown in Fig. 3.12. As the particle factory was set on the leftmost side of the figure, it can be seen from the red part on the left side of the figure that the particle has the maximum speed when it fell initially. As it fell on the scraper conveyor and moves forward with the scraper chain, the state and speed of particles tended to be stable gradually. In addition, according to the velocity nephogram, the velocity of the coal

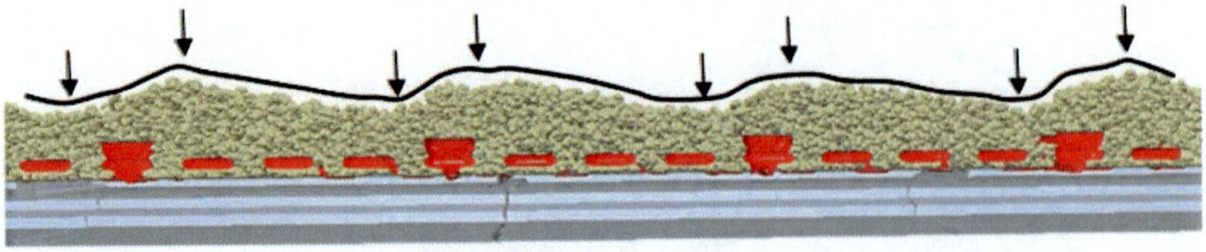

Fig. 3.11 Longitudinal section of middle trough

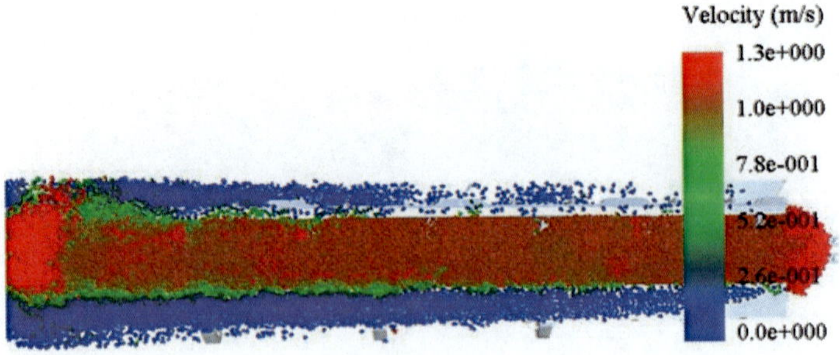

Fig. 3.12 Velocity nephogram of coal bulk in Y direction

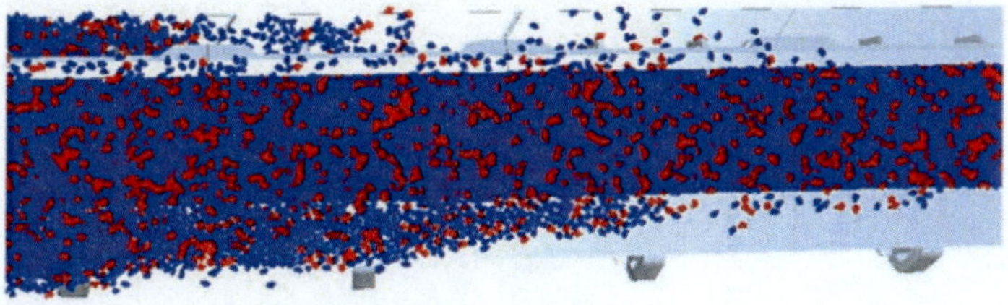

Fig. 3.13 Distribution of coal bulk and gangue particles

bulk in the middle part of the middle trough is larger than that near the sides of the middle trough. When the particles fall on the middle trough, the particles tend to be stationary, as shown in the blue part of the figure.

The distribution of coal bulk and gangue particles on the middle trough is shown in Fig. 3.13, in which the blue part represents coal bulk and the red part represents gangue. It can be seen from the figure that the distribution of gangue and coal bulk on the surface of the middle trough presents a random state.

3.2.4 Bottom Plate Tilting Condition

The influence of the bottom plate tilt on the coal bulk transportation was analyzed by setting the middle trough tilt in the EDEM. The bottom plate was inclined 5° clockwise and anticlockwise, respectively. In the simulation, the working parameters were, respectively, set as: chain speed 1.1 m/s, conveying capacity 330 kg/s and simulation 13 s.

Figure 3.14 shows the distribution of coal bulk in the middle trough when the bottom plate was inclined 5° clockwise and 5° anticlockwise, respectively. It can be seen that when inclined clockwise, the coal bulk are stacked toward the right side. When tilted counterclockwise, the bulk is stacked to the left. When there is too much accumulation, coal bulk will be scattered from the slot side of the accumulation side.

The three curves in Fig. 3.15 show the average mass flow rate changes when the bottom plate of the middle trough is horizontal, inclined clockwise and anticlockwise, respectively. The average mass flow rate here represents the average mass of coal bulk passing through a length of middle trough in unit time, which can be used to evaluate the transportation efficiency of the middle trough. It can be seen from the curve change that when the bottom plate is horizontal, the value of the average mass flow stability period is larger, which is 120 kg/s. When the bottom plate is inclined, the fluctuation of the average mass flow becomes larger, and the value of the stable period is smaller than that of the bottom plate level, which is 110 kg/s.

According to the particle distribution pattern and average mass flow curve, when the bottom plate is inclined and the middle trough is under full load operation, there will be particles scattered from the slope side of the slot. In addition, the incline will make the transportation efficiency of the conveyor worse.

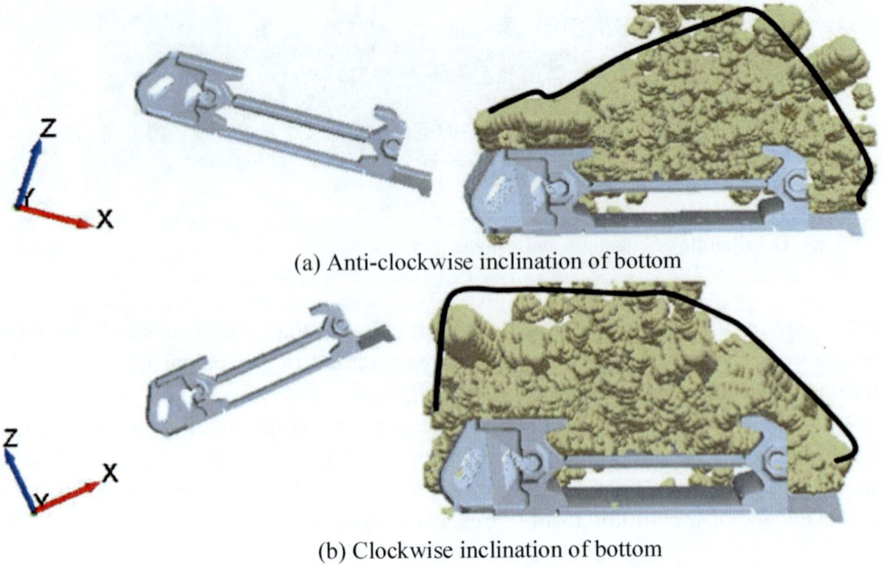

Fig. 3.14 Distribution pattern of coal bulk with 5°

Fig. 3.15 Average mass flow rate curve of middle trough

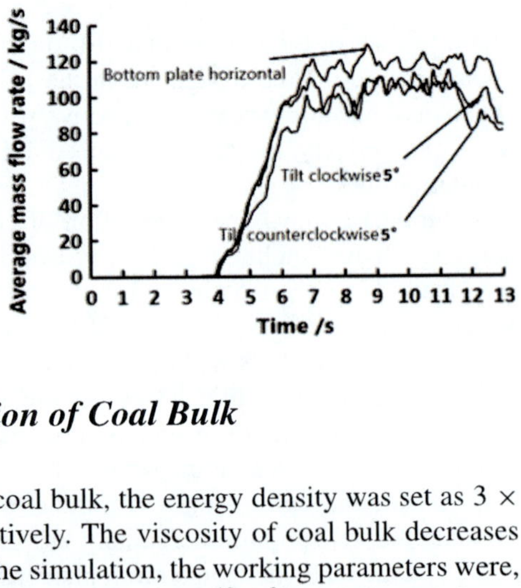

3.2.5 *Viscosity Change Condition of Coal Bulk*

According to the different viscosity of coal bulk, the energy density was set as 3×10^6, 4×10^6 and 5×10^6 J/m^3, respectively. The viscosity of coal bulk decreases with the decrease of energy density. In the simulation, the working parameters were, respectively, set as follows: wear constant was 1.5×10^{-12} m^2/N, chain speed was 1.1 m/s, conveying capacity was set as 330 kg/s and simulation time was 13 s.

The distribution of coal bulk with different viscosity is shown in Fig. 3.16. When the energy density was 3×10^6 J/m^3 (Fig. 3.16a), the particles scattered outside the middle trough were not bound into blocks. However, with the increase of energy density (Fig. 3.16b), the coal bulk were caked and agglomerated at 4×10^6 J/m^3. The larger the energy density (5×10^6 J/m^3), the larger was the volume of the

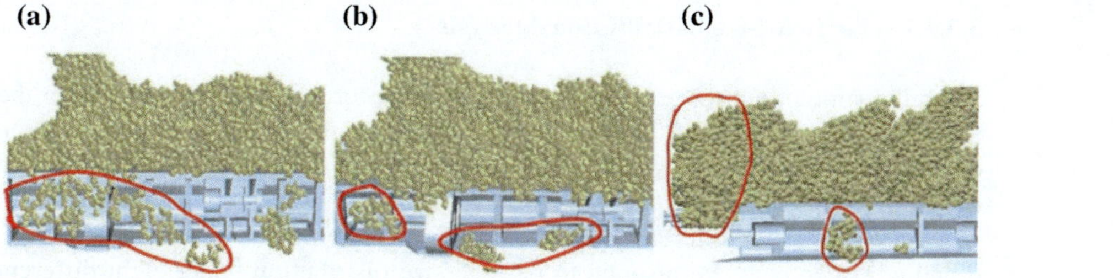

Fig. 3.16 The distribution pattern of coal bulk with different viscosities

bonded block. As shown in Fig. 3.16c, the coal bulk are bonded into a large block and suspended at the end of the middle trough. The analysis shows that the larger the energy density is, that is, the higher water moisture makes the viscosity larger, the more likely it is for the middle trough to bond and block during transportation. When the coal bulk are smaller, the phenomenon of coalescence becomes more serious with the increase of energy density.

3.3 Distribution and Movement State of Coal Bulk

3.3.1 Distribution State of Coal Bulk

In this section, on the basis of the above simulation of various working conditions, combined with the simulation results obtained, the particle state of coal bulk will be analyzed and studied. It includes particle size distribution, velocity distribution, velocity change and other movement characteristics of coal bulk, as well as the whole movement state of coal bulk under different working conditions, so as to provide reference for relevant research and structural design optimization of scraper conveyor.

In order to conform to the actual situation as much as possible, the multi-circle filling method was used to construct the coal bulk. The standard particle model is composed of four spherical surfaces with a radius of 40 mm. To reflect the influence of different particle sizes on the transport characteristics of coal bulk, three different particle sizes were used for simulation in this part. The three particle sizes are 0.5, 1 and 1.3 times of the standard particle size, respectively, and they are generated randomly by the particle factory in the simulation process to ensure that the generated particles fall into the middle trough of the scraper conveyor in a mixed state.

3.3.1.1 Particle Size Distribution Analysis

After the simulation, the particles were colored according to the mass size in the post-processing of EDEM. The simulation results can clearly reflect the distribution of particles with different sizes. The particle size distribution under various stable conditions is shown in Fig. 3.17.

Figure 3.17 is the reference model simulation test profile. The four groups b, c, d and e, respectively, correspond to particle size distribution under eight different conditions, including two different mining heights, two different feeding speeds, two different transportation angles and two different transportation speeds. It can be seen from the figure that under various working conditions, the large particles represented by red are basically concentrated in the upper part of the bulk, while the small particles represented by blue are basically concentrated in the lower part of the bulk and the medium particles represented by green are basically concentrated in the middle part of the bulk. It shows that in the actual coal transportation process, the particle size distribution of the stable coal bulk is basically independent of the mining height, feeding speed, transportation angle, transportation speed and other parameters. All of them show the distribution state of big block on top and small block at the bottom.

3.3.1.2 Analysis of Speed Distribution

The previous simulation results were used. Under the color setting subtemplate of the 'Post-processing' module (Analyst), the min value and max value under the attribute coloring tab were set to 1.09 and 1.1 m/s, respectively. The particle coloring interval was set to 1.09–1.1 m/s, the speed was colored blue, green and red in order from small to large, and then different positions were selected to longitudinally cut the middle trough of the scraper conveyor to obtain the particle speed distribution longitudinally. The sectional view is shown in Fig. 3.18.

Figure 3.18a and b is the velocity profile obtained by cutting the middle trough of scraper conveyor. The former is the longitudinal profile near the scraper area, and the latter is the longitudinal profile between the two scraper areas. The velocity distribution of the particles near the scraper was slightly different. The velocity of the particles near the bottom of the middle trough is larger than that near the bottom of the middle trough. From Fig. 3.18c, it is obvious that the particle velocity distribution in the middle trough is not in a continuous state, but shows regional concentration, and the center of concentration is scraper. This clarifies that only the particles in the front part of the scraper have the same velocity as that of the scraper, while the other particle velocities are lower than that of the scraper.

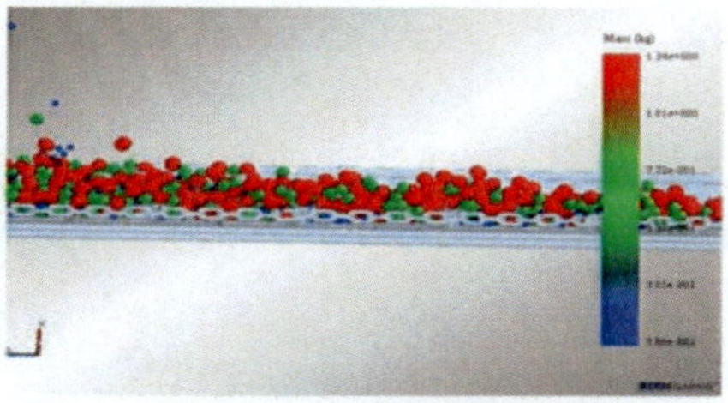

(a) Reference model simulation

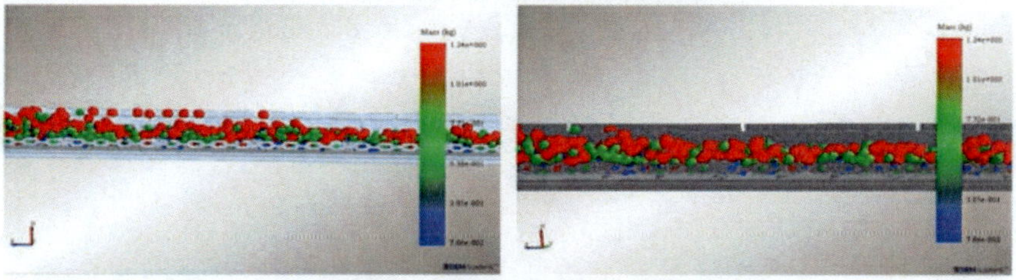

(b) Simulation of different mining height models

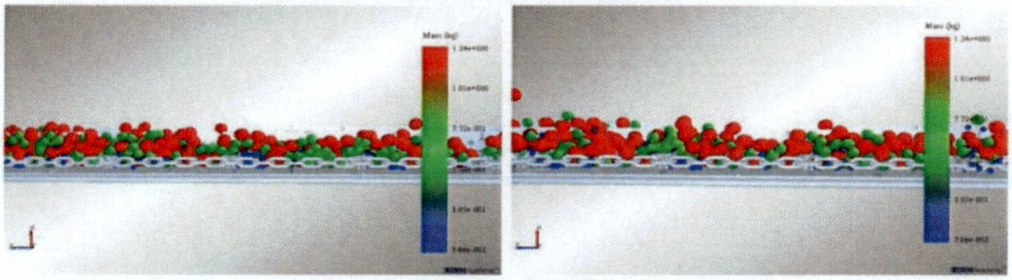

(c) Simulation of different feeding speed models

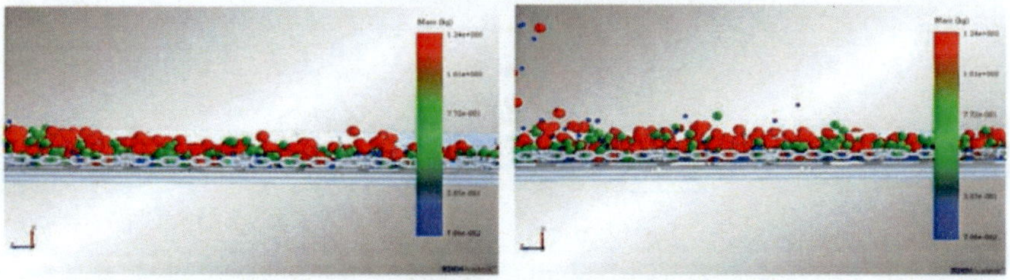

(d) Simulation of different transport angle models

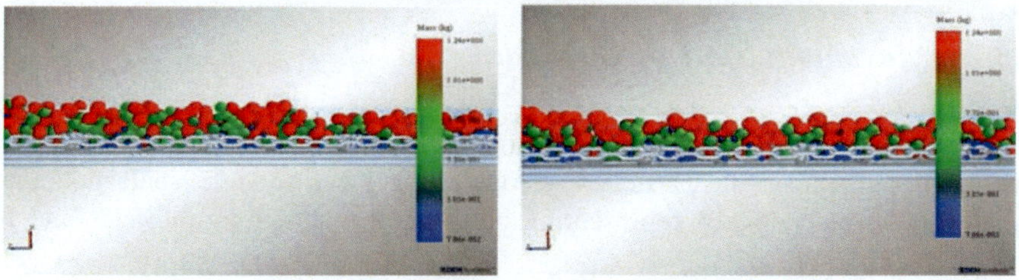

(e) Simulation of different transport speed models

Fig. 3.17 Particle size distribution profile under different working conditions

Fig. 3.18 Coal bulk velocity distribution (coloring interval 1.09–1.1 m/s)

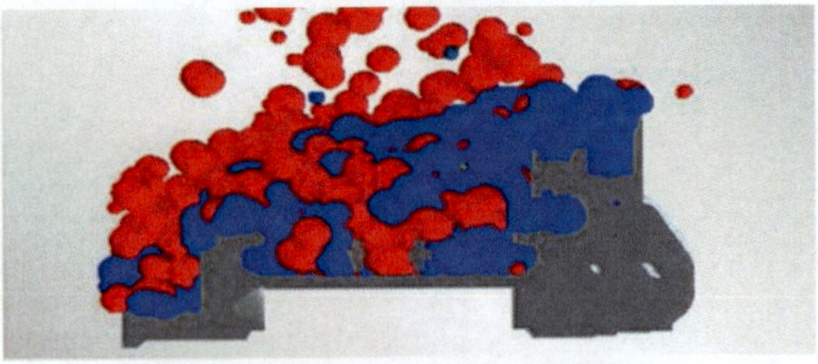

(a) Velocity distribution profile

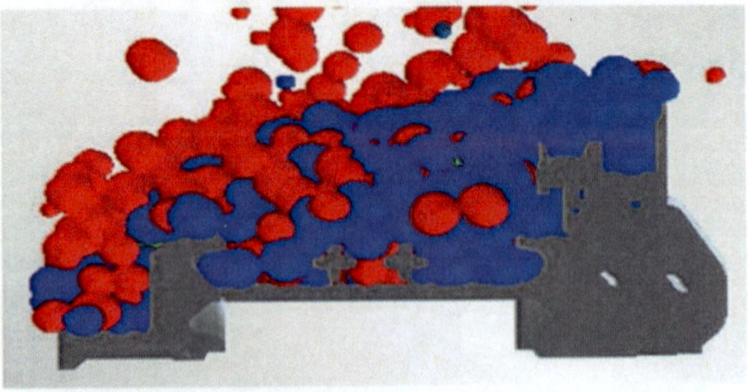

(b) Velocity distribution

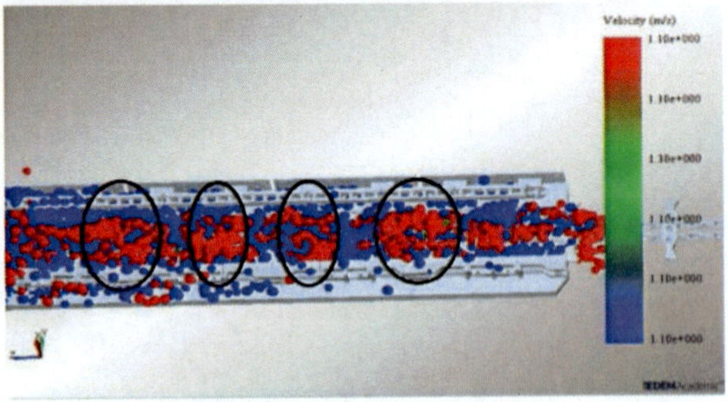

(c) Speed zone concentration

3.3.2 *Movement State of Coal Bulk*

As shown in Fig. 3.19, the coal particle model was re-filled with spheres to represent three different shapes and particle sizes, including 1-ball model, 2-ball model and 3-ball model, respectively, representing ball-like particles, long particles and flat particles.

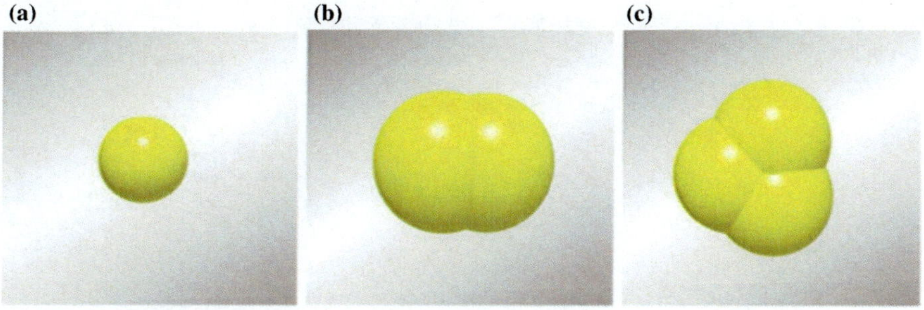

Fig. 3.19 Coal particle model

3.3.2.1 Coal Bulk Flow Pattern

In the model subinterface, set the opacity to 0 to hide the 3D solid of the conveyor. The display mode of particles was set as stream, which was represented by different colors, among which blue, green and red represent particles 1, 2 and 3, respectively. The streamline direction was set as the speed direction, the length was the speed size and the step length was 0.3 s. Then the flow state curve of coal bulk was obtained by playing. The influence of different coefficient of static friction between coal and steel on flow state was studied. Three kinds of settings set in the simulation were: $\mu = 0.4$, $\mu = 0.6$ and $\mu = 0.8$. Figure 3.20 shows the flow under different coefficients of static friction at 7 s. It can be seen from the observation and analysis that as the coal bulk fall into the middle trough, the flow state (speed size and direction) tended to be stable gradually; the coefficient of static friction between coal and steel had an obvious influence on the flow of coal bulk, and the greater the μ, the greater is the uncertainty of the speed direction of bulk particles. Therefore, the analysis shows that when the coefficient of static friction between coal and steel is too large, it is not conducive to the transportation of coal bulk.

3.3.2.2 Coal Bulk Velocity

(1) Change of speed with time

Randomly select one particle from the ball particle, strip particle and flat particle for tracking. The curve chart of speed of the particles in horizontal direction (Z-axis) with time was derived, as shown in Fig. 3.21. The analysis shows that the speed changes of different particle types on the conveyor are basically similar, which shows that the speed tends to be stable from vibration, and the final speed is stable near the scraper chain speed with small amplitude fluctuation.

Fig. 3.20 Flow pattern of coal bulk at 7 s

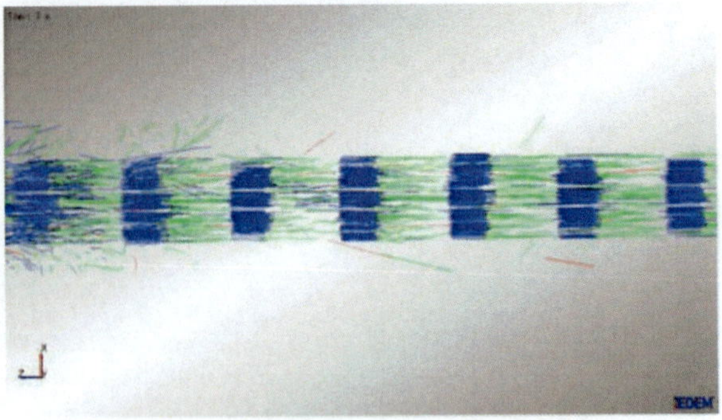

(a) μ=0.4

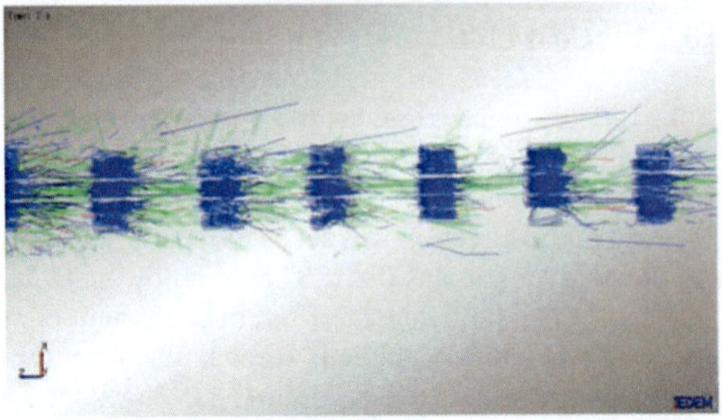

(b) μ=0.6

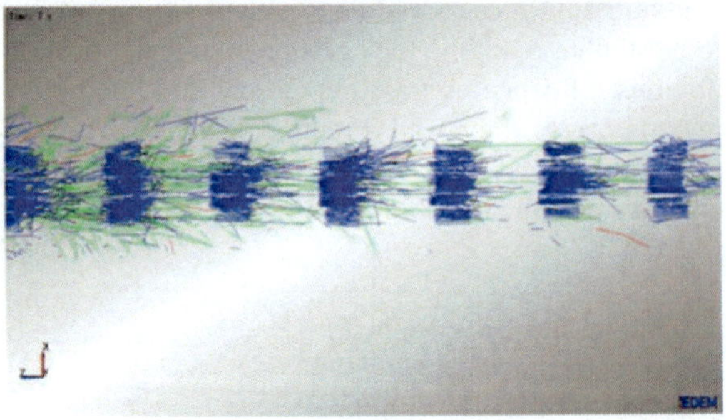

(c) μ=0.8

(2) Velocity distribution area

In the model interface of the post-processing module of EDEM, the display mode of the middle trough and scraper chain was set to mesh, the opacity was 0.05, the display mode of particles was vector and the shading attribute of particles was set to speed. At this time, the size, direction and color of the vector, respectively, represented the mass size, velocity direction and velocity size of the particles. According to the velocity vector diagram (Fig. 3.22) of the particle group at 8 s, the whole particle

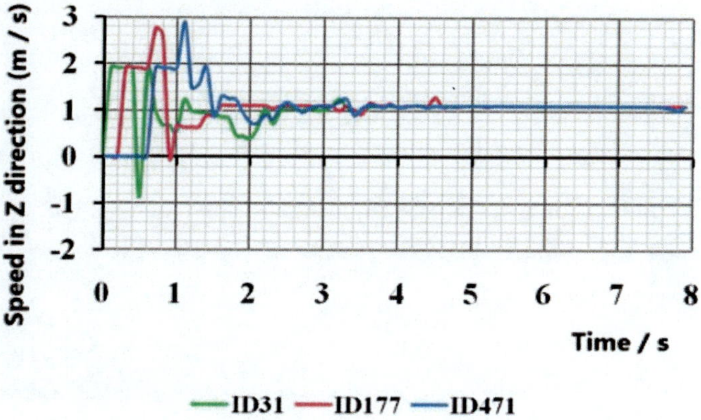

(a) The change curve of Z-direction velocity of particle 1 with time

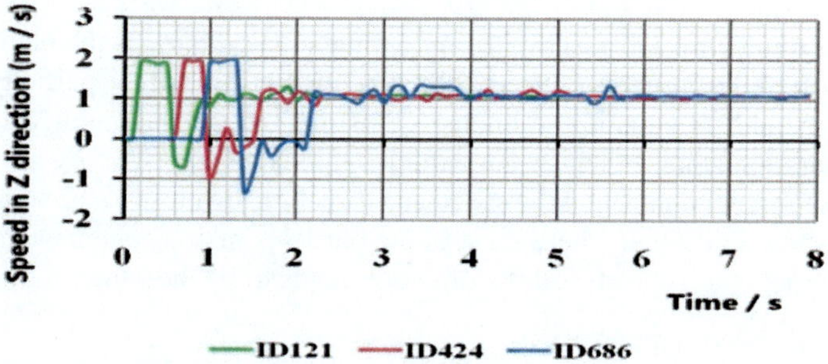

(b) The change curve of Z-direction velocity of particle 2 with time

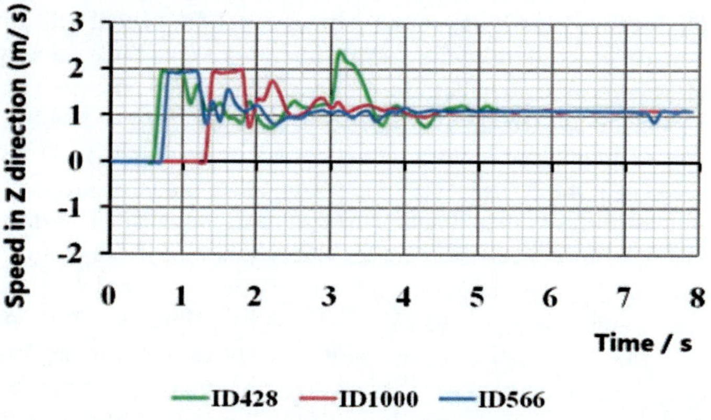

(c) The change curve of Z-direction velocity of particle 3 with time

Fig. 3.21 Change curve of coal bulk velocity in Z direction with time

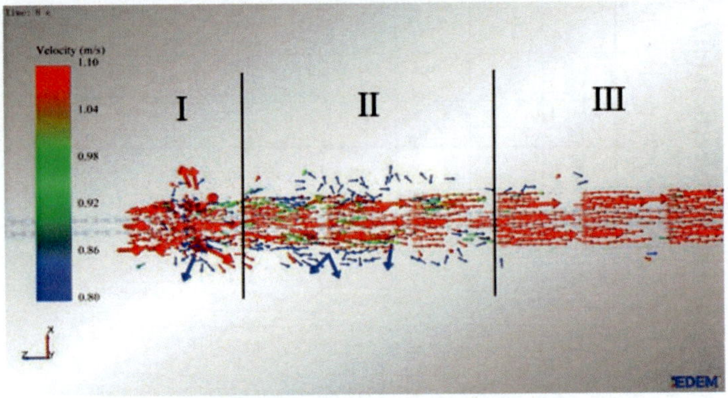

Fig. 3.22 Speed vector diagram at 8 s

group in the middle trough was divided into three areas: Zone I is the blanking area, in which most of the particles fall vertically downward, some of the particles rebound at the bottom, and the situation of intense collision is the most complicated. Zone II is the acceleration zone, in which the particles in this zone move forward gradually driven by the scraper chain, and the speed and direction are close to the scraper chain. Zone III is the stable zone, where most of the particles in this zone have been stable near the scraper chain speed, and the direction tends to be the same.

(3) Change of speed with position

Figures 3.23 and 3.24 show the scatter diagram of the change of coal bulk velocity with position. At 6, 7 and 8 s, the velocity scatters in the Z direction (horizontal transportation direction) and Y direction (vertical direction) of coal bulk group were, respectively, made. According to the scatter diagram of coal bulk velocity in the Z direction, the velocity of coal bulk in the horizontal direction changes greatly when it just falls into the middle trough of scraper conveyor. After the coal bulk move with the scraper chain, their speed tends to the chain speed gradually, and only shows a small fluctuation near the scraper (periodic particle accumulation area in Fig. 3.23). It can be seen from the scatter diagram of coal bulk velocity in Y direction that the vertical velocity of coal bulk increases linearly before it falls into the middle trough. (Figure 3.23 shows the inclined linear particle accumulation area.) After falling into the middle trough, the velocity in the vertical direction fluctuates around 0, the amplitude of the oscillation decreases after its forward movement driven by the scraper chain and the velocity in the vertical direction gradually decreases until it is reduced to 0, and only the scraper nearby will show a small amplitude of fluctuation (periodic particle accumulation area in Fig. 3.24).

Fig. 3.23 Velocity scatter diagram in the Z direction

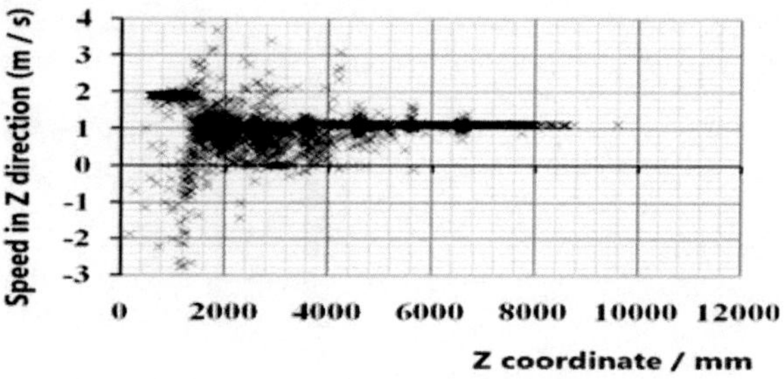

(a) At 6s

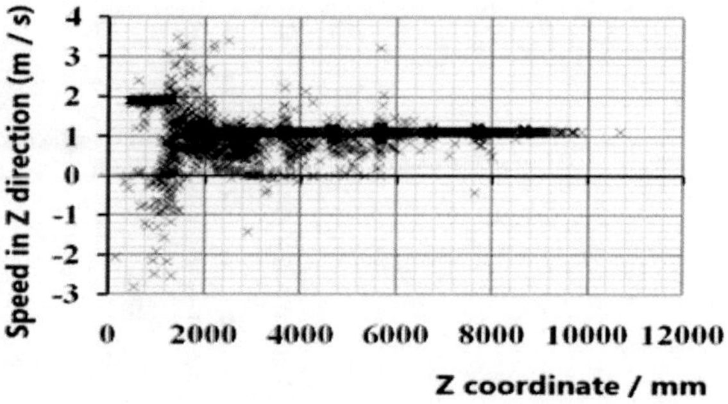

(b) At 7s

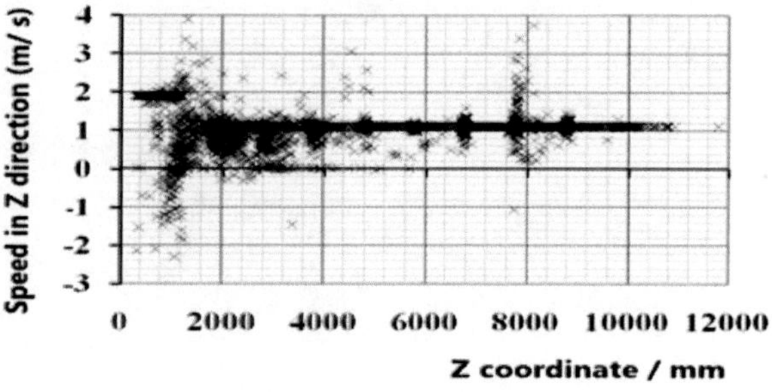

(c) At 8s

Fig. 3.24 Velocity scatter diagram in the Y direction

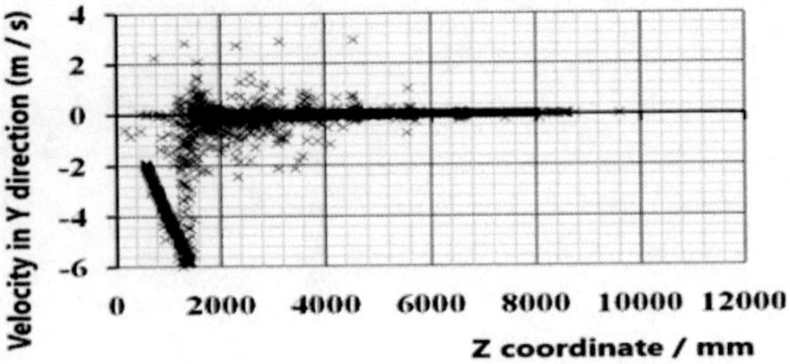

(a) At 6 s

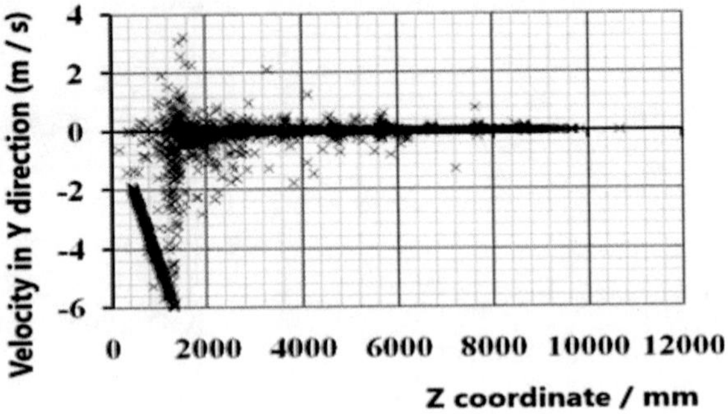

(b) At 7s

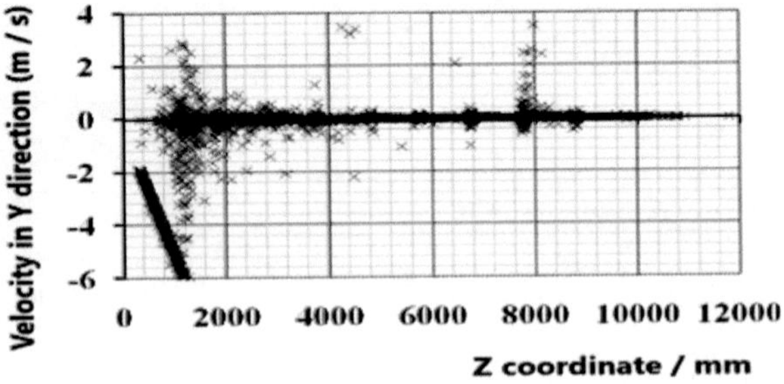

(c) At 8s

Reference

1. Li T, Wang X, Li B, Li J, Yang Z (2018) Optimization method for coal particle model parameters based on discrete element method. China Powder Technol 24(5):6–12 (in Chinese)

Chapter 4
Mechanical Effect of Scraper Conveyor Based on DEM

4.1 EDEM Simulation Analysis of Bulk Load Characteristics

4.1.1 Reconstruction of Scraper Conveyor Model

4.1.1.1 Coupling Model

Only using EDEM to simulate cannot accurately describe the real movement state of chain link or scraper, and the scraper and chain link directly imported into EDEM will not be affected by coal bulk or other geometry changes, and the mechanical relationship between the coal bulk and the middle trough is not completely accurate. RecurDyn has the advantage in complex geometry kinematics and dynamics. Coupling EDEM with RecurDyn can make the scraper and chain obtain the freedom in real situation and realize the simulation of the real state of coal bulk conveyed by scraper conveyor.

4.1.1.2 Coal Particle Model

It can be seen from Fig. 4.1 that the shape of the coal particle model was established in EDEM. Figure 4.1a is quasi block; Fig. 4.1b is flat; and Fig. 4.1c is quasi cone. The particle size of the three shapes was set as 40 mm, and the generation rate of coal bulk was set as 200 kg/s.

X. Wang et al., *Engineering Applications of Discrete Element Method*, Engineering Applications of Computational Methods 4,
https://doi.org/10.1007/978-981-15-7977-6_4

Fig. 4.1 Particle model shape

4.1.2 Load Characteristics of Coal Bulk Under Normal Working Conditions

In this section, by simulating the running state of the middle trough under normal working conditions in which the slope of uphill was 0°, the compression force distribution of coal in the trough was studied as follows.

The load distribution is characterized by the pressure of coal bulk, and the stable transportation state of coal bulk in the middle trough at 5 s, which is selected to draw the internal pressure nephogram of coal bulk. As shown in Fig. 4.2a, it is a cloud chart of bulk pressure cutting along a plane perpendicular to the X axis, and cutting along the e-e line in Fig. 4.2c. The red coal bulk is mainly located in the inner side of the slot in front of each scraper, and the coal bulk above the slot is mainly blue-green. It can be seen that the coal bulk pressure in the middle trough is significantly greater than the coal bulk pressure accumulated in the upper part of the slot. In Fig. 4.2b, A, B and C are mainly red; D, E and F are mainly blue; B is less red than A and C; and E is bluer than D and F. Therefore, in the Y direction, the bulk pressures in A, B and C regions are respectively greater than those in D, E and F regions. In the X direction, the pressure of coal bulk in B area is less than A and C, and that in E area is less than D and F. The average pressure of the six areas at 5 s is shown in Table 4.1.

The pressure of the coal bulk in the upper part of the ledge is greater than that of the coal bulk in the upper part of the ledge. The coal bulk in the ledge is squeezed by scraper, middle plate and ledge, and moves forward under the action of extruding force. The coal bulk piled up in the upper part of the ledge is moved forward by the friction force exerted by the coal bulk at the bottom, so the pressure of the coal bulk is obviously smaller than that of the inner part of the ledge.

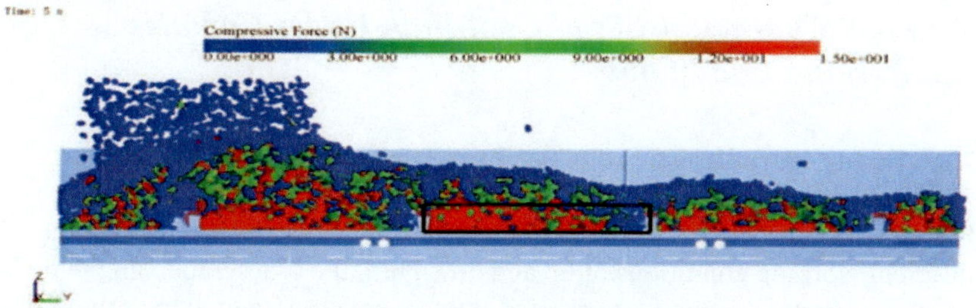

(a) Plane cut perpendicular to X axis

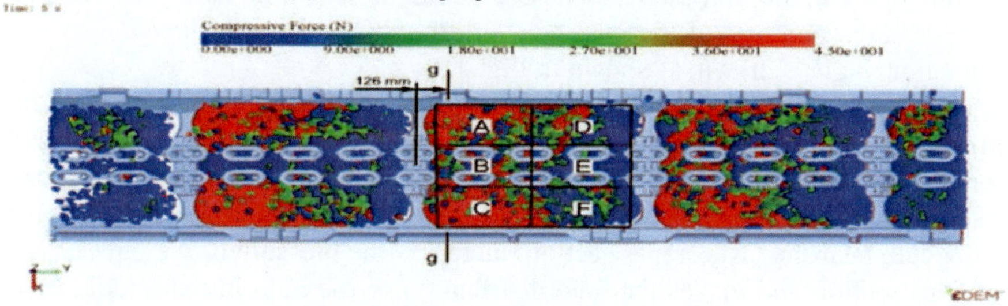

(b) Plane cut perpendicular to Z axis

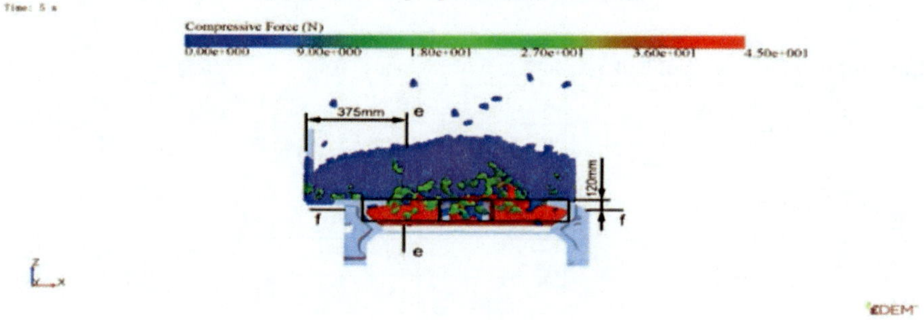

(c) Plane cut perpendicular to Y axis

Fig. 4.2 Nephogram of coal bulk pressure at 5 s

Table 4.1 Average pressure of bulk in six areas at 5 s

Region	A	B	C	D	E	F
Average compression (F)/N	87	68	84	19	11	16

4.1.3 *Load Characteristics of Coal Bulk Under Complex Working Conditions*

4.1.3.1 Load Characteristics of Coal Bulk Under Up and Down Hill Conditions

For different working conditions of up and down hill, the two scrapers in the middle trough of the second section were divided into six areas at 5 s to extract the average pressure of coal bulk in four areas A, B, D and E, as shown in Table 4.2. The curve of the average pressure of the coal bulk in different areas with the change of the inclination angle is drawn, as shown in Fig. 4.3.

The gravity of the uphill condition intensifies the trend of the coal bulk approaching to the rear scraper, increases the bulk pressure (area A, B) near the rear scraper and reduces the bulk pressure (area D, E) near the previous scraper. On the contrary, in downhill condition, the effect of gravity reduces the pressure of the bulk near the last scraper section, increases the pressure of the bulk near the previous section and makes the load distribution of the bulk in the middle trough more uniform.

Table 4.2 Average value of bulk pressure in four regions at 5 s

Up and down slope angle/°	A/N	B/N	D/N	E/N
5	174	79	20	11
0	85	35	18	13
−5	137	24	12	8
−10	72	16	14	5
−15	56	15	18	8

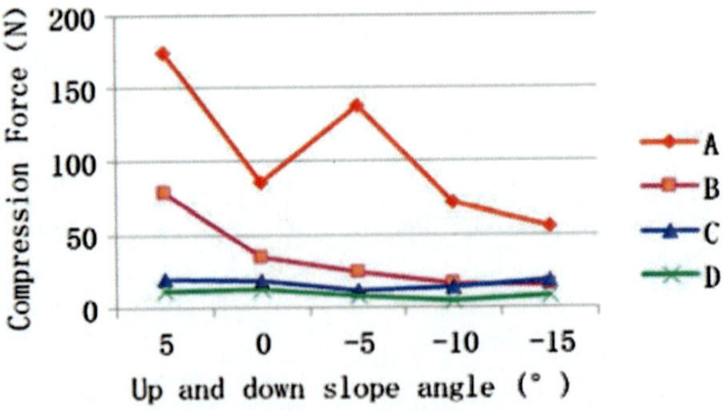

Fig. 4.3 The pressure variation of coal bulk in each region under up and down hill conditions

4.1.3.2 Inclined Condition of Middle Trough Along the Moving Direction

Under the condition that the middle trough inclines 10° along the moving direction, the pressure nephogram of the coal bulk in the ledge is drawn, as shown in Fig. 4.4, which is the top view of the pressure nephogram of the coal bulk cut in the middle of the ledge in a plane perpendicular to the Z-axis. The red particles on the right side of the chain are obviously more than those on the left, indicating that the pressure of coal bulk on the right side of the chain is greater than that on the left side. The areas A and C inside the two ledges were selected and fixed with the scraper behind. The average value of the bulk pressure of the two areas A and C was calculated when the middle trough is inclined by 5° and 0° and the value is shown in Table 4.3. F_C/F_A = 86% at 5° and 96% at 0°, indicating that with the increase of the inclined angle of the middle trough, the uneven degree of the pressure on both sides of the chain in the ledge increases.

The inclination of the middle trough causes the coal bulk to accumulate to one side of the ledge, which makes the load distribution of the coal bulk in the ledge uneven. The load on the coal bulk on the inclined side of the ledge is the largest in the area between the two scraper sections close to the two sides of the chain of the latter scraper section. This kind of load distribution of coal bulk will lead to uneven load on both ends of scraper and affect the normal movement of scraper.

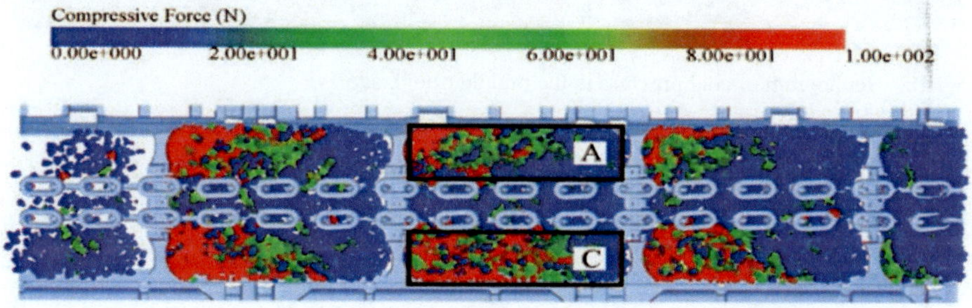

Fig. 4.4 Nephogram of the pressure of coal bulk in the chute

Table 4.3 The average pressure and its ratio

Slope angle of ledge/°	0	5	10
F_A/N	102	119	83
F_C/N	106	138	118
F_A/F_C	96%	86%	70%

4.1.3.3 Different Transport Capacity Conditions

This section analyzes the relationship between coal load and transportation capacity. Figure 4.5 is the pressure nephogram of the coal bulk in the middle trough at 5 s under the transportation capacity of 400 kg/s. As shown in the figure, from right to left, the number of red coal bulk in front of the three scrapers increases gradually. In order to determine the quantitative relationship between the transportation capacity and the maximum load of the coal bulk in the slot top, the fixed area S in the ledge in Fig. 4.5 is selected to extract the average pressure of the coal bulk. At 2.1, 3.1, 4.1 and 5.1 s, the average pressure and mass flow rate data of coal bulk in area S under three capacity conditions are extracted, as shown in Table 4.4.

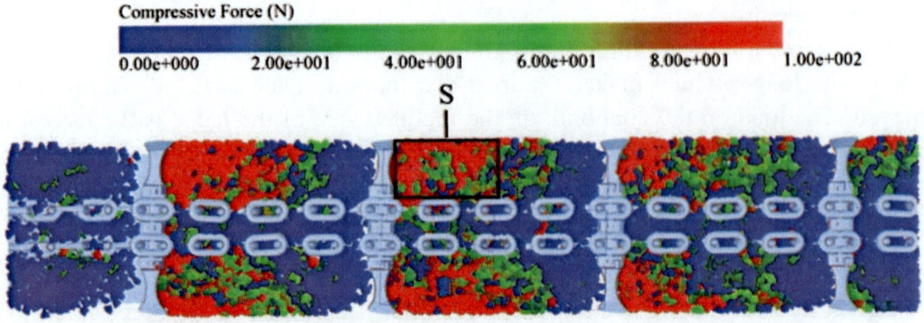

Fig. 4.5 Nephogram of coal pressure in the middle trough at 5 s

Table 4.4 Data of mass flow rate and average pressure of coal in region S

Mass flow rate/kg/s	Average pressure/N
152	36
169	75
180	54
193	56
194	63
233	77
245	119
246	84
247	96
257	93
282	124

4.1.3.4 Local Stacking Condition of Ledge

As shown in Fig. 4.6, the pressure of the coal bulk in the ledge increases significantly at 3.5 s. As shown in Fig. 4.7, the pressure data of coal bulk in areas A, B and C are extracted to draw the change curve of pressure. The sudden accumulation of coal bulk makes the fluctuation peak of bulk pressure in A and C areas to suddenly increase, and then slowly decrease to the normal value. The peak points of periodic change of bulk pressure in area C are selected, and the curve of peak point is fitted out, which approximately expresses the trend of bulk load change in the slot.

The load change characteristics of the coal bulk in the ledge are related to the accumulation and flow of the coal bulk. When the coal bulk suddenly accumulates on one side of the slot, the impact force is exerted on the coal bulk being transported. At the same time, the accumulated coal bulk also increases the conveying capacity of the coal bulk in the middle trough. In addition, the accumulated coal gradually flows into the ledge and is transported away with the decrease of the coal in the ledge, so the coal load in the ledge increases at the moment of accumulation, and then decreases gradually. The periodical change of coal load is not always great, but the periodical peak value decreases gradually.

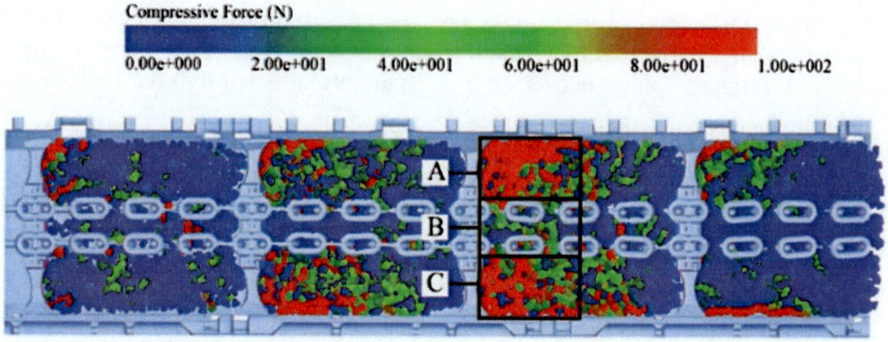

Fig. 4.6 Nephogram of pressure of coal bulk in the chute at 3.5 s

Fig. 4.7 Variation curve of pressure of coal bulk in the chute under coal accumulation condition

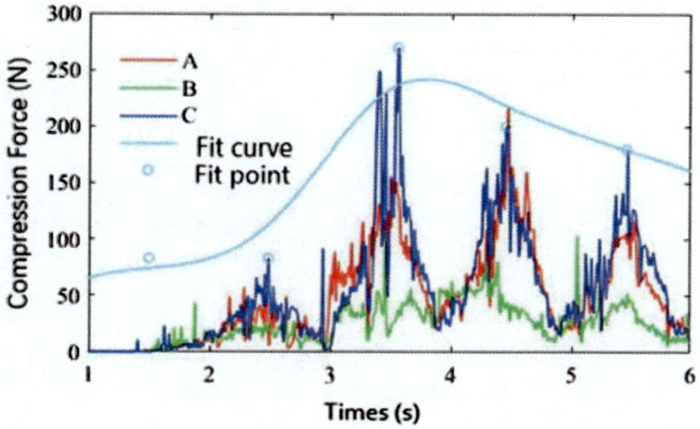

4.2 EDEM Simulation Analysis of the Force Features of the Middle Trough of the Scraper Conveyor

As the main component of the scraper conveyor, the force characteristics of the middle trough have a great influence on its own life, and the force characteristics of the middle trough-related components are analyzed from different operating conditions.

4.2.1 Up and Down Hill Conditions

4.2.1.1 The Load of Coal to the Middle Plate

The middle plate is the component used to carry coal in the middle trough; the load mainly comes from the action of the scraper and the chain. The middle plate is fixed in the middle trough, which does not move with the scraper and the coal, the load of the middle plate is related to the coal, scraper and chain in the middle trough and the force of the other fixed parts of the middle trough can refer to the force characteristics of the middle plate. Figure 4.8 is a diagram of the position of the scraper, chain and middle plate at the end of the simulation. There is no direct contact between the middle plate, the scraper and the chain, so the scraper and the chain have no direct force on the middle plate, but its force is indirectly affected by coal bulk as the medium.

Combined with the load characteristics of coal bulk in the preceding section, the second section of the middle trough is selected to analyze the force characteristics of the middle plate. Figure 4.9 is the second middle trough at 6 s. The middle plate is

Fig. 4.8 Position relation diagram of scraper and middle plate

Fig. 4.9 Schematic diagram of middle plate area selection

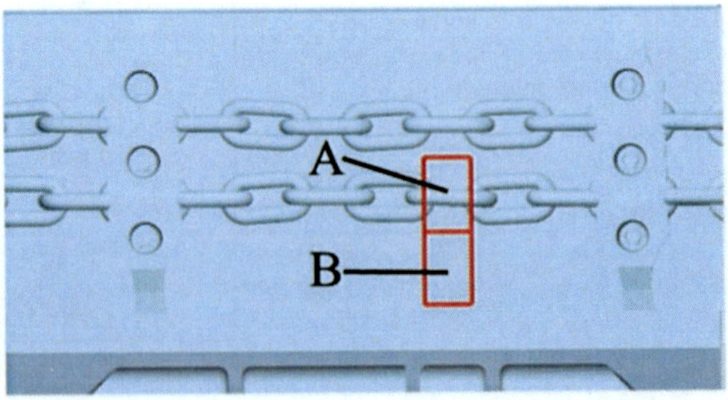

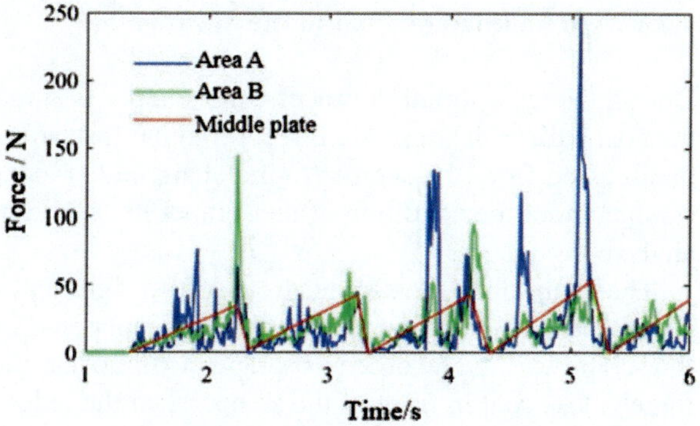

Fig. 4.10 Load curve of local area of the middle plate

symmetrical to the center axis with the intermediate dividing line of the two chains, so the areas A and B are selected for the middle plate force analysis. Area A is located just below the chain, area B is in the middle position of the slot and the chain, the force curve of the middle plate along the coal feed of the two areas is shown in Fig. 4.10 and the red curve is the trend curve of the force of the middle plate in area A and B. It can be seen that the changing trend of the magnitude of the forces in the two areas has a high similarity in general, and the change law is basically consistent, showing a certain periodic change. In contrast, however, the volatility of the force magnitude in area B is more moderate and more regular. This is because, under the influence of the chain force, the coal bulk between the lower part of the chain and the middle plate in area A exert a greater load on the middle plate, and the middle plate force is therefore mutated. From the analysis it can be seen that the change of the force of the middle plate is related to the scraper plate to promote the coal movement, taking a scraper and coal bulk pushed by it for a unit analysis, and the force of the middle plate changes periodically with time based on the time that a unit passes through the fixed area of the middle plate. The force at the end of each cycle is relatively large or sometimes suddenly increases to a large; this is because the load of coal bulk deposited in front of the scraper is large, and sometimes the coal bulk will be caught between the scraper and the middle plate, so that the force of the middle plate increases.

The influence of the up and down hill condition on the middle plate is similar to the effect on the coal bulk load, and the periodic variation of the bulk load shows a decreasing trend with the increase of the angle of the downhill, and distribution between the two sections of the middle trough is also more uniform.

4.2.1.2 The Load of Coal to the Scraper

During the transportation process, the scraper is subjected to resistance applied by the coal bulk in the opposite direction to the forward direction, and the resistance is then passed from the scraper to the chain and sprocket. The coal resistance of the scraper under the condition of the scraper under different angles of the mountain is analyzed.

The chain pulls the scraper from left to right under the drive of the sprocket to push the coal. According to the order of scraper entering the groove to push the coal, the scraper will be recorded as scraper 1, scraper 2 and scraper 3. At the beginning, there is less coal in front of the scraper 1 in the ledge, so choose a relatively stable scraper 2 to analyze its force. Due to the existence of the pre-tension force of the chain itself and the action of external coal resistance, the chain pull force is much greater than the coal resistance to the scraper. Influenced by the polygon effect of chain drive, the fluctuation of chain pull has a strong periodicity, with a period of about 0.25 s. For the coal resistance curve, as the scraper begins to push the coal forward movement, the resistance of coal to scraper increases rapidly. When the coal fully moved, the resistance of the scraper reaches the maximum, and then the resistance gradually decreases to a certain level during the movement. It fluctuates up and down after that, and eventually decreases to 0° as the scraper leaves the ledge. Comparing the three curves, it is found that the resistance curve and the pull curve have a clear positive correlation, and their volatility is very similar, especially in the stable stage of resistance fluctuation. Influenced by the polygon effect of chain drive, the scraper movement has obvious periodicity, which causes the interaction force between coal bulk and scraper to fluctuate with the same characteristics (Fig. 4.11).

Under the condition of going up and down hill, the gravity of coal has different effects affected by different inclination angles, which hinders the movement of coal when going uphill and promotes the movement of coal when going downhill, thus affecting the resistance of scraper. Because the particle model in the simulation does not occur in the reality of coal fragmentation phenomenon, then coal stuck between the middle plate and the scraper will have a much greater force than the actual value. In order to avoid the effect of this accidental situation leading to sudden change of load, compare the resistance of the scraper under several inclination conditions to obtain the normal resistance of scraper 1 and scraper 3 (Fig. 4.12). Overall, the resistance to scraper with the increase of the angle of the downhill gradually decreases. From the uphill to the downhill, the gravity component of coal in the transportation direction is gradually transformed from resistance to power, so the resistance to the scraper gradually decreases. It can be seen from Fig. 4.13 that the resistance of the scraper under the downhill working condition is smaller, especially under the downhill condition with 15°, the resistance to the scraper is only 65% of condition that the uphill angle is 0°, which greatly reduces the power and cost of equipment.

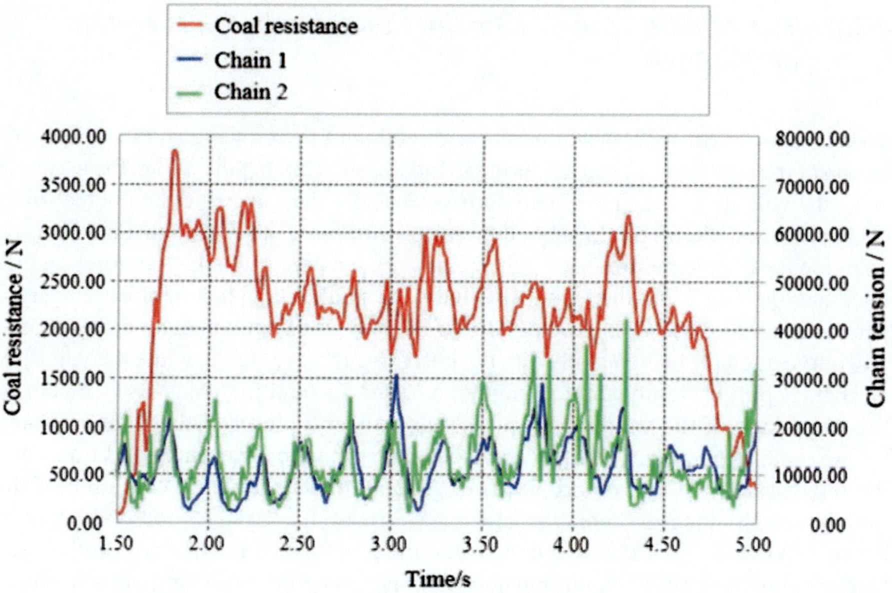

Fig. 4.11 Chain tension and scraper 2 resistance curve by the coal

Fig. 4.12 The average resistance curve of scraper with different inclinations angles

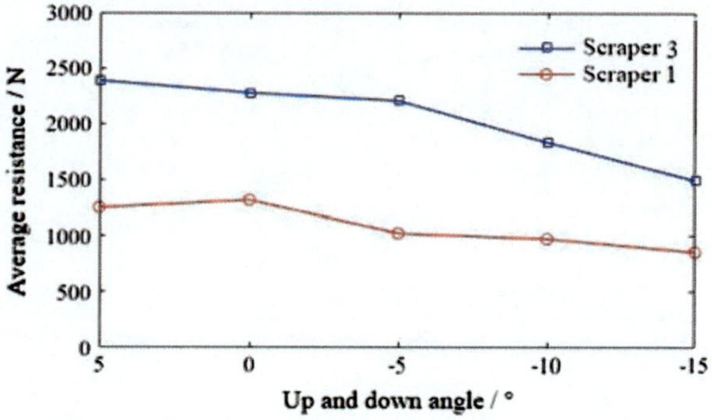

Fig. 4.13 The coal resistance to scraper percentage of different working conditions to zero degree working conditions

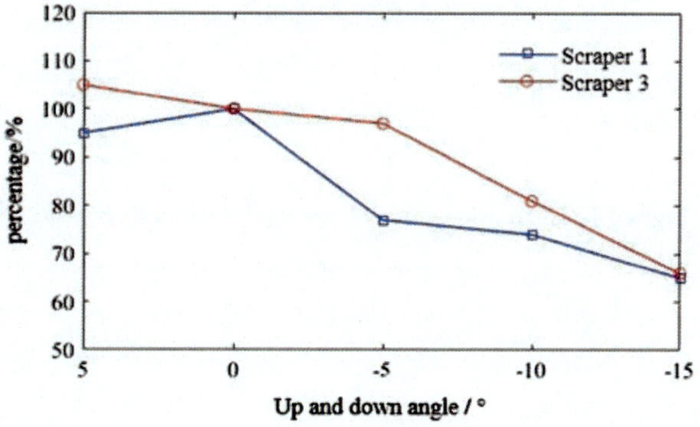

4.2.2 *The Middle Trough Tilts the Condition in the Direction of the Push*

When the middle trough was tilted in the direction of the movement, the load distribution of the coal was uneven, resulting in inconsistent chain pull. When the tilt angle is 0°, the volatility of the two chain pulls is basically the same, and the fluctuation is cyclical. When the tilt angle is 5°, the tension curves of the two chains have a high coincidence degree before 2.5 s. After 2.5 s, the tension of chain 1 is more volatile than that of chain 2, but the tension of chain 1 is smaller than that of chain 2 overall. When the tilt angle is 10°, the volatility of the two chains pull is quite different; the pull curve of chain 1 is relatively flat, the pull force is more stable, while the volatility of chain 2 pull is significantly strengthened after 3 s, and the volatility is higher to the larger side of the value. After 4.5 s rear scraper 2 closes to the right sprocket, the two chain pull curves show a clear trend under the three working conditions, and the fluctuation is more synchronized. In general, under the tilting condition of the middle trough along the direction of the movement, the match of the pull curve of the two chains is significantly reduced, and the pull force is not uniform. Influenced by the polygon effect of chain transmission, the closer the two chain pulls from the sprocket, the stronger the synchronization of fluctuations (Fig. 4.14).

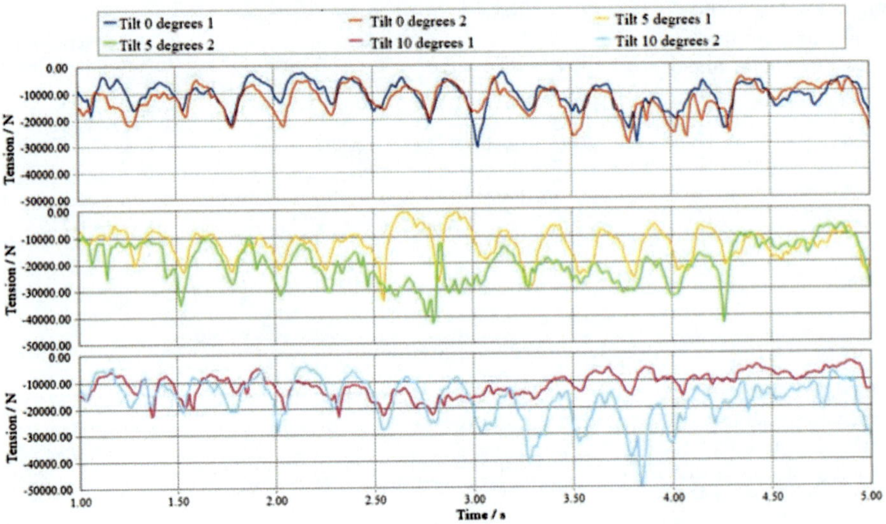

Fig. 4.14 The tension curve of two chains under three working conditions

4.2.3 Different Capacity Conditions

With the increase of coal transportation volume, the resistance of the scraper to the coal bulk gradually increases, but the concrete relationship within is still to be examined. According to the sequence in which four scrapers enter the ledge to convey coal, scrapers were recorded as scraper 1, scraper 2, scraper 3 and scraper 4. Overall, in the coal bulk transport process, the peak of the four scrapers affected by the resistance of coal is increasing, and the first peak generates at about 1 s (Fig. 4.15). In order to reveal the relationship between the resistance of the scraper and the coal quality flow rate, starting from the coal resistance exerted on the scraper, the resistance of the scraper within 1.5 s after extraction was taken and averaged (Table 4.5). The process of fitting the coal mass flow rate with the scraper resistance curve was used in MATLAB (Fig. 4.16). The fitting curve is a quadratic function curve, and the analytical formula is $y = 3.3162x^2 - 128.8x + 15680$ ($R^2 = 0.994$). It can be seen that the higher the coal quality flow rate, the greater the scraper resistance, and the growth rate is faster and faster.

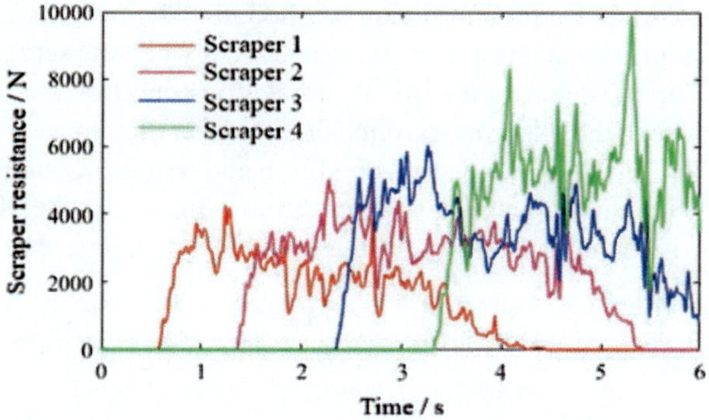

Fig. 4.15 Coal resistance curve of the scraper

Table 4.5 Data of scraper resistance and coal mass flow rate

Scraper	Scraper resistance/N	Coal mass flow rate/kg/s
Scraper 1	2598	200
Scraper 2	3171	250
Scraper 3	4063	270
Scraper 4	4884	290

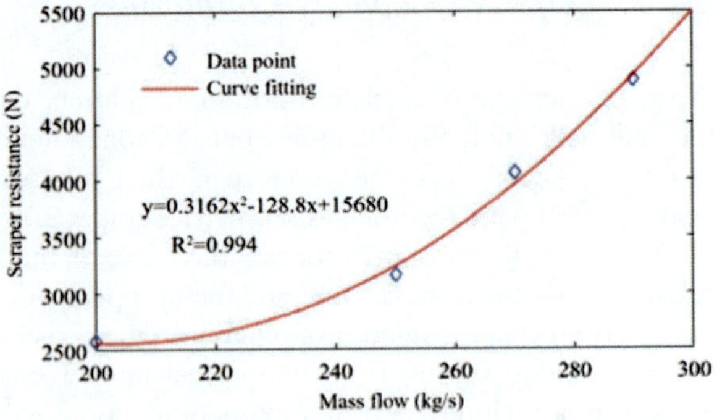

Fig. 4.16 Curve of relation between coal mass flow rate and scraper resistance

4.2.4 Ledge Local Accumulation Conditions

The local accumulation of coal bulk has a large impact effect on the scraper and chain in operation, and the force of the scraper and chain is compared and analyzed from the three conditions of no accumulation, accumulation of 200 and 400 kg. It can be seen from Fig. 4.17 that at 1.5 s, scraper 2 participates in coal conveying, and at about 3 s, the accumulated coal begins to fall into the middle trough until scraper 2 disengages from the ledge at about 5.5 s. The three curves before 3 s basically keep changing synchronously, and the resistance of the scraper fluctuates back to a certain range after being washed up. The non-accumulation condition keeps this state until the resistance is reduced to 0 when the scraper is separated from the ledge. The resistance of the two accumulation conditions suddenly increases around 3 s and rose to the maximum around 3.25 s. After that, the resistance decreases rapidly under the 200 kg accumulation condition and remains basically synchronized with the non-accumulation condition after 3.5 s, while under the 400 kg accumulation condition where the resistance fluctuates steadily for a period of time, the resistance gradually

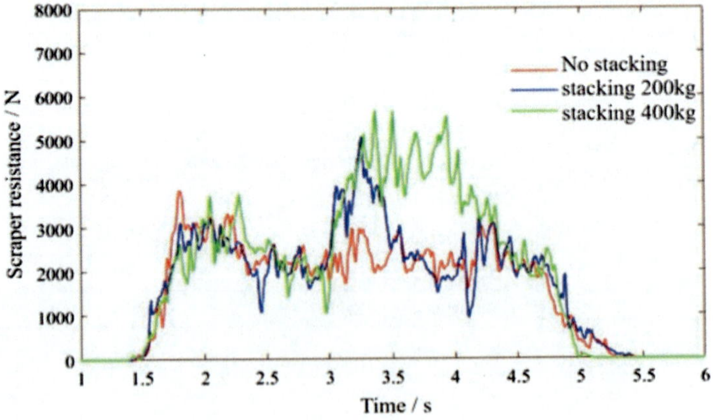

Fig. 4.17 Resistance curve of coal bulk to scraper under different accumulations

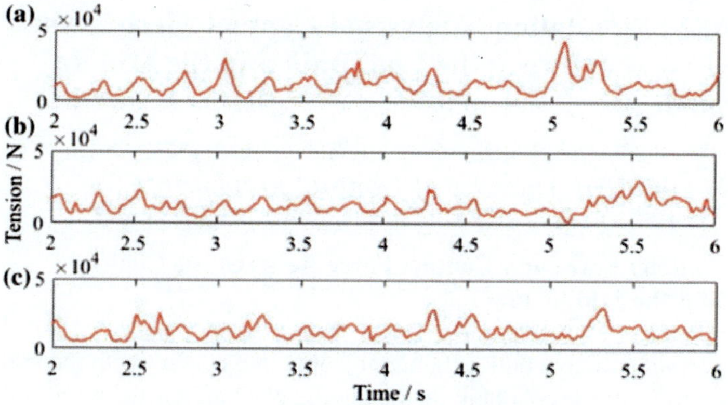

Fig. 4.18 Various curve of chain tension under different accumulations

decreases until the scraper leaves the ledge and drops to 0. The local accumulation of coal bulk produces a large impact, resulting in a sharp rise in the resistance of the scraper to the coal, but whether the resistance can be maintained at a high level is determined by the quality of the accumulated coal. And the greater the mass, the stronger the impact, the longer the resistance remains at a high level, but the resistance will eventually fall back, and remains in sync with the non-accumulation conditions.

Figure 4.18 is the pull change curve of the chain ring connected to the scraper 2 under three operating conditions: the pull curve when there is no accumulation; the pull curve when the stack is 200 kg; and the pull curve when the stack is 400 kg. According to the characteristics of the chain transmission, along the direction of the chain movement, the increase of the front chain pull will not affect the pull of the rear chain, but the increase of the pull of the rear chain will cause the front chain pull to become larger. As previously known, the accumulated coal starts to fall into the ledge around 3 s. It can be found from the figure that under no stacking condition, chain pull change has a significant periodicity, and the polygon effect of chain transmission has a significant influence on the chain pull. The impact of the material greatly weakens the periodic change of chain pull, and the chain tension changes suddenly around 3.25, 4.25 and 5.25 s, and the mutation interval is approximately 1 s, which is equivalent to the time interval between the adjacent scrapers passing through the fixed area of the middle plate.

Because the chain pull is much larger than the resistance of the scraper to the coal, the impact effect of the coal local accumulation condition has no significant effect on the overall magnitude of the chain pull, but it weakens the fluctuation of the chain pull, and the degree of weakening is related to the quality of the accumulated coal: the greater the quality, the stronger the weakening effect, and when the amount of coal accumulation is larger, the chain pull will have a mutation with the scraper through the coal accumulation.

4.3 EDEM Simulation Analysis of Contact Mechanics and Wear Between the Coal Bulk and the Middle Trough

4.3.1 Theoretical Analysis of Contact Mechanics

4.3.1.1 Contact Form and Contact Force Between the Coal and the Middle Plate

The transportation of coal bulk is bound to come into contact with the middle plate. The contact force of the coal bulk on the middle plate affects the wear of the middle plate. Therefore, in order to analyze the contact force between the particles with different shapes and the middle plate, as well as the stress–strain characteristics of the middle plate, in the contact between the coal and the middle plate, the coal is regarded as a rigid body, and the middle plate is regarded as an elastic half-plane. The contact between the particles and the middle plate can be simplified as the contact between the cylinder, sphere, cone rigid body and elastic half-plane [1].

(1) Deformation of elastic half-space body under normal and tangential surface force

(a) Deformation of elastic half-space body under normal surface force

As shown in Fig. 4.19, a concentrated force acts on the elastic half-space body. According to the relevant knowledge of elastic mechanics, the deformation of the elastic half-space body can be calculated. When $z = 0$, the deformation displacements of the surface of the elastic half space in three directions are:

$$\begin{cases} u_x = -\dfrac{(1+\nu)(1-2\nu)}{2\pi E}\dfrac{x}{r^2}F_z \\ u_y = -\dfrac{(1+\nu)(1-2\nu)}{2\pi E}\dfrac{y}{r^2}F_z \\ u_z = \dfrac{\left(1-\nu^2\right)}{\pi E}\dfrac{1}{r}F_z \\ r = \sqrt{x^2+y^2} \end{cases} \tag{4.1}$$

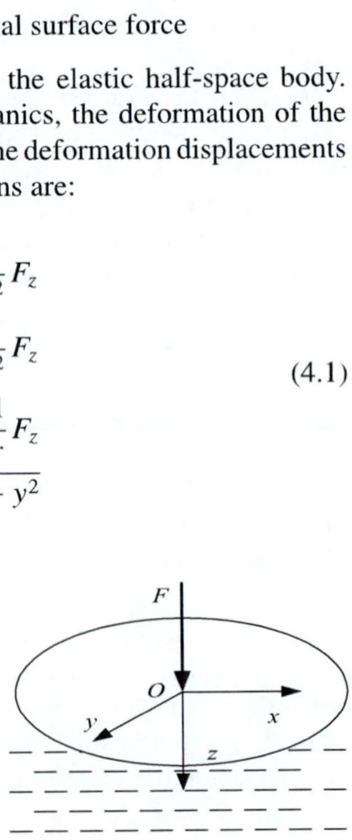

Fig. 4.19 The normal concentration force acting on the elastic half-plane

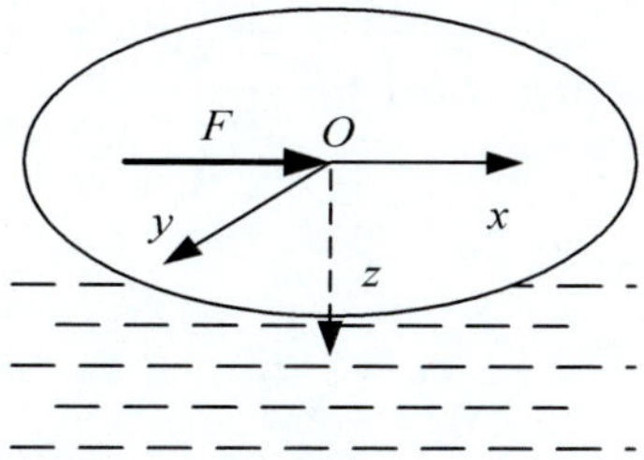

Fig. 4.20 The tangential concentrated force is acted on the elastic half-plane

(b) Deformation of elastic half-space body under the action of tangential surface force

As shown in Fig. 4.20, a tangential concentrated force is acted at a point on the surface of the elastic half-space body. The point of action is taken as the origin and the direction of the force is the x-axis. Then the deformation equation of the elastic surface (z = 0) is:

$$\begin{cases} u_x = F_x \dfrac{1}{4\pi G}\left[2(1-\nu) + \dfrac{2\nu x^2}{r^2}\right]\dfrac{1}{r} \\ u_y = F_x \dfrac{1}{4\pi G} \cdot \dfrac{2\nu}{r^3} xy \\ u_z = F_x \dfrac{1}{4\pi G} \cdot \dfrac{1-2\nu}{r^2} x \end{cases} \tag{4.2}$$

where G is the shear modulus.

(2) The contact force generated by the three shapes of particles

(a) Contact between cylindrical rigid body and elastic half-plane (Fig. 4.21)

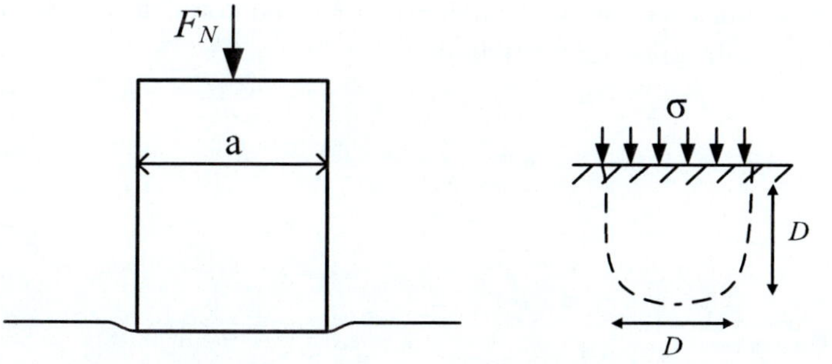

(a) Contact diagram (b) Severely deformed area

Fig. 4.21 The contact between cylinder and elastic half-space body

$$p = p_0\left(1 - \frac{r^2}{a^2}\right)^{-1/2} \tag{4.3}$$

where

$$p_0 = \frac{1}{\pi}E^*\frac{d}{a} \tag{4.4}$$

$$E^* = \frac{E}{1 - \nu^2} \tag{4.5}$$

p_0 is the maximum pressure, E is the elastic modulus of the material and v is the Poisson's ratio.

The vertical displacements of all points in the contact area are equal, and the vertical displacement component of the severely deformed area of the elastic half-space body along the z-axis is:

$$u_z = \frac{\pi p_0 a}{E^*} \tag{4.6}$$

The contact force generated by the contact between the two is:

$$F_N = 2\pi p_0 a^2 = 2aE^*d = cd \tag{4.7}$$

where c is the contact stiffness, $c = 2aE^*$.

(b) Contact between sphere rigid body and elastic half-plane

As shown in Fig. 4.22, a rigid sphere with radius R is in contact with an elastic half-plane, and the indentation depth of sphere is d. Contact radius a meets:

$$a^2 = Rd \tag{4.8}$$

The pressure distribution on the contact surface conforms to the Hertz pressure distribution, and the pressure distribution is:

$$p = p_0\left(1 - \frac{r^2}{a^2}\right)^{1/2} \tag{4.9}$$

where

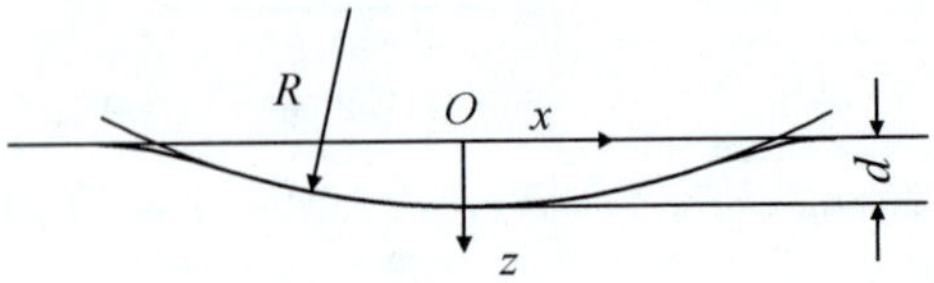

Fig. 4.22 The contact between the sphere rigid body and the elastic half-plane body

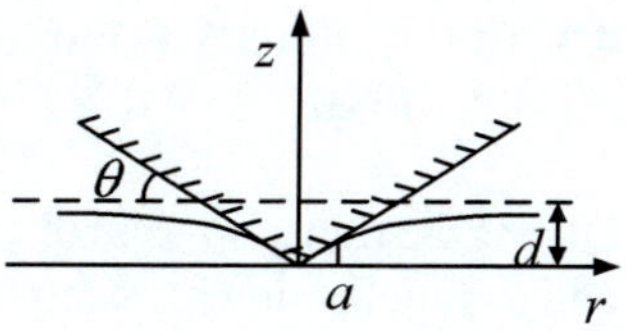

Fig. 4.23 The contact between the vertebral body and the elastic half-plane

$$p_0 = \frac{2}{\pi} E^* (\frac{d}{R})^{1/2} \tag{4.10}$$

The vertical displacement component of an elastic half-space body under stress is:

$$u_x = \frac{1}{E^*} \frac{\pi p_0}{4a} (2a^2 - r^2) \tag{4.11}$$

The contact force generated by the contact between the two is:

$$F_N = \frac{2}{3} p_0 \pi a^2 = \frac{4}{3} E^* R^{1/2} d^{3/2} = cd \tag{4.12}$$

where

$$c = \frac{4}{3} a E^* = \frac{4}{3} (Rd)^{1/2} E^* \tag{4.13}$$

(c) Contact between cone rigid body and elastic half-plane

As shown in Fig. 4.23, a rigid cone is pressed into the elastic half-space body. The relationship between the indentation depth d of the cone and the contact radius a is:

$$d = \frac{\pi}{2} a \tan\theta \tag{4.14}$$

The expression of pressure distribution is:

$$p(r) = -\frac{Ed}{\pi a(1 - \nu^2)} \ln\left(\frac{a}{r} + \sqrt{\left(\frac{a}{r}\right)^2 - 1}\right) \tag{4.15}$$

The contact force is:

$$F = \frac{2}{\pi} E^* \frac{d^2}{\tan\theta} \tag{4.16}$$

4.3.1.2 Contact Mechanical Effect Between the Coal and the Middle Plate

During the transportation of coal in the middle trough, there are mainly sliding friction and rolling friction between the coal bulk and the middle plate. Both kinds of friction have the contact and force of coal and the middle plate. The contact force between them seriously affects the friction and wear effect of the middle plate. These two kinds of friction are called contact mechanical effects that occur between them. This section theoretically analyzes the force characteristics of sliding and rolling friction between a rigid sphere and an elastic half-space body.

(1) Sliding contact

When there is a tangential force between two objects in contact, the two objects tend to move relative to each other. According to Coulomb's law of friction, when the tangential force is greater than the frictional force, the two objects slide relative to each other, causing sliding friction. Sliding friction is divided into two stages, when $\tau < \mu p$, local slip occurs, and when $\tau > \mu p$, complete slip occurs.

When $\tau < \mu p$, the circular tangential contact area is divided into an adhesive area and a sliding area, as shown in Fig. 4.24a, then the stress distribution is given by

$$\tau = \tau^{(1)} + \tau^{(2)} \tag{4.17}$$

where

$$\begin{cases} \tau^{(1)} = \tau_1 (1 - \frac{r^2}{a^2})^{-1/2} \\ \tau_1 = \mu p_0 \end{cases} \tag{4.18}$$

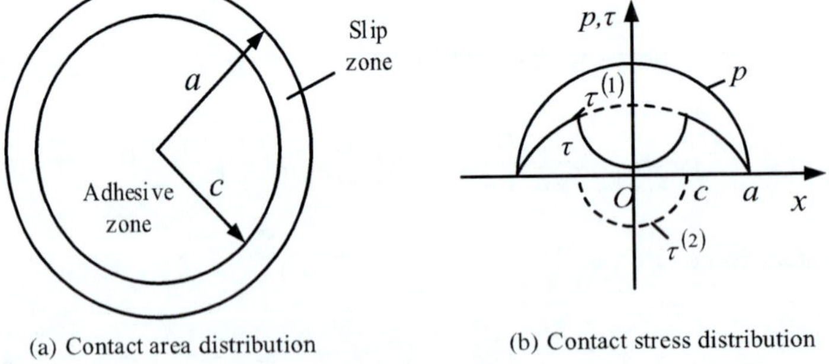

Fig. 4.24 The sliding contact area and stress distribution

$$\begin{cases} \tau^{(2)} = -\tau_2(1 - \frac{r^2}{c^2})^{-1/2} \\ \tau_2 = \mu p_0 \frac{c}{a} \end{cases} \tag{4.19}$$

The stress distribution is shown in Fig. 4.24b.

The deformation displacement along the x-axis under the action of tangential force is:

$$u_x = \frac{\tau_1 \pi}{32Ga}\left[4(2-\nu)a^2 - (4-3\nu)x^2 - (4-\nu)y^2\right] - \frac{\tau_2 \pi}{32Gc}\left[4(2-\nu)c^2 - (4-3\nu)x^2 - (4-\nu)y^2\right] \tag{4.20}$$

where the displacement in the adhesion zone is constant.

$$u_x = \frac{(2-\nu)\pi \mu p_0}{8Ga}\left(a^2 - c^2\right) \tag{4.21}$$

The tangential contact force is:

$$F_x = \frac{2\pi}{3a}\mu p_0\left(a^3 - c^3\right) = \mu F_N\left[1 - (\frac{c}{a})^3\right] \tag{4.22}$$

The radius of the adhesion zone is:

$$\frac{c}{a} = \left(1 - \frac{F_x}{\mu F_N}\right)^{1/3} \tag{4.23}$$

When $\tau > \mu p$, complete sliding occurs. This time,

$$F_x = \mu F_N \tag{4.24}$$

(2) Rolling contact

Rolling friction is similar to local sliding friction, and there are both adhesive and sliding areas. According to Carter theory, the stress distribution of rolling contact is constructed by superimposing Hertz pressure distribution. As shown in Fig. 4.25a, it is the distribution diagram of the adhesion area and the rolling area of the rigid sphere rolling on the elastic half-space body. According to the characteristics of rolling, there is always an adhesion area on the leading edge of the rolling, and the trailing edge is the sliding area. The stress distribution in the contact area is:

$$\tau = \tau^{(1)}(x) + \tau^{(2)}(x) \tag{4.25}$$

where

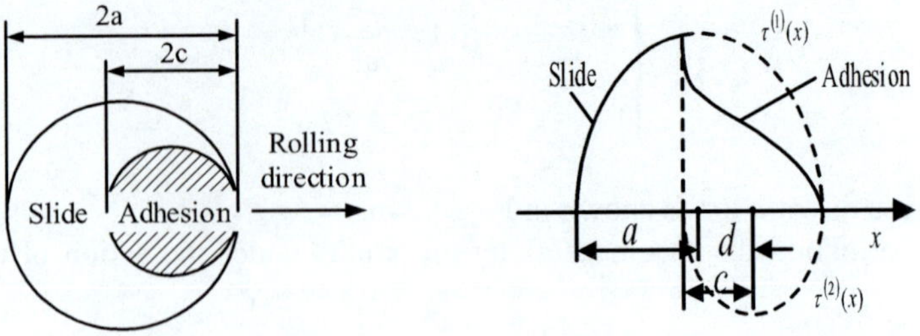

(a) Contact area distribution (b) Contact stress distribution

Fig. 4.25 The rolling contact area and stress distribution

$$\tau^{(1)}(x) = \tau_1\sqrt{1 - \frac{x^2 + y^2}{a^2}} \tag{4.26}$$

$$\tau^{(2)}(x) = -\tau_2\sqrt{1 - \frac{(x - d)^2 + y^2}{c^2}} \tag{4.27}$$

And τ only exists in the adhesion area. The stress distribution curve along the x-axis of the center of the contact area is shown in Fig. 4.25b.

The surface deformation due to superimposed stress is:

$$u_x = \frac{\pi}{32G}\left\{\begin{aligned} &\frac{\tau_1}{a}4[(2 - \nu)a^2 - (4 - 3\nu)x^2 - (4 - \nu)y^2] \\ -&\frac{\tau_2}{c}[4(2 - \nu)c^2 - (4 - 3\nu)(x - d)^2 - (4 - \nu)y^2 \end{aligned}\right. \tag{4.28}$$

The tangential force is:

$$F_x = \frac{2}{3}\pi a^2\tau_1 - \frac{2}{3}\pi c^2\tau_2 = \mu F_N\left[1 - (\frac{c}{a})^3\right] \tag{4.29}$$

The radius of the adhesion area is:

$$\frac{c}{a} = \left(1 - \frac{F_x}{\mu F_N}\right)^{1/3} \tag{4.30}$$

4.3.2 Analysis of the Wear Mechanism of the Middle Trough

The wear of the middle trough is mainly caused by coal bulk being furrowed and micro-cutting the surface of the middle trough under the action of a scraper (chain), which is three-body abrasive wear; when the coal bulk are directly in contact with the middle trough, it has a scraping effect on the surface of middle trough, and also causes wear on the middle trough, which is two-body abrasive particles. Based on this, it is judged that the wear mechanism of the middle trough is mainly abrasive wear, and the specific form is: the coal bulk is not subject to the force of the scraper and the chain during the transportation process, and the wear caused by the coal bulk acting on the middle trough is the two-body wear; coal bulk is subjected to the force of the scraper and chain and then act on the middle plate. The wear on the middle trough is three-body wear.

The chain is formed by the interlocking connection of flat and vertical rings, forming a concave and convex space between the two chains. The coal bulk is filled in it. During the transportation process, the two chains move synchronously to drive the coal bulk between them to move forward together. The movement of the chains presents the polygonal effect of the chain transmission, and the coal bulk between them is affected by the pressure and also shows the periodic change of the polygonal effect of the chain transmission. The pressure of coal bulk on both sides of the chain path is also affected by the movement of the chain, but it is only affected by one chain. As the distance between the coal bulk and the chain increases, its influence gradually disappears, so it is mainly affected by the accumulation shape of the coal bulk in the middle trough.

The most important factor for the wear of the middle plate caused by coal bulk is that the normal contact force between the coal bulk and the middle plate is greatly increased by the abnormal load, which causes serious wear to the middle plate. There are two situations: coal bulk are stuck between the scraper and the middle plate and between the vertical chain and the middle plate. Due to the cumulative effect, the coal bulk are stuck between the chain and the middle plate, which causes more serious wear to the middle plate. This is why the most serious wear occurs in the middle plate chain-path. The other areas of the middle plate can be divided into two areas according to the degree of wear: between the chain-paths and on both sides of the chain-paths, and the wear between the chain-paths is more serious. In both areas, the load on the middle plate is periodically changing. Although the average load on both sides of the chain-path is greater, the period of load fluctuation between the chain-paths is 1/4 of that on both sides of the chain path. The load of the coal on the middle plate is a periodic fluctuating load, so that the wear of the coal on the middle plate has a saw tooth effect. The friction of the coal on the middle plate like saw tooth intensifies the wear of the middle plate, and the smaller the load cycle is, the more severe the wear caused to the middle plate is, relative to the denser the saw tooth.

4.3.3 Analysis of the Wear Law of the Middle Trough

The physical properties of coal bulk include Poisson's ratio, shear modulus, density and hardness, and these physical properties of coal vary greatly between different types of coal. The Poisson's ratio of coal reflects the lateral deformation coefficient of coal under unidirectional tension or compression, generally between 0.26 and 0.42.

The shear modulus of coal represents the coal's ability to resist shear strain. A large shear modulus indicates that the coal is strong. And the shear modulus of coal is generally between 1×10^8 and 9×10^8 Pa. The density of coal is also very different because of the difference of coal type and mining area, generally in the range of 1100 ~ 1900 kg/m^3. The hardness of coal is mainly related to the type of coal. The Mohs hardness of coal is between 2 and 4. In the theoretical calculation, the hardness of coal is reflected by the wear constant, and the wear constant between the coal and steel contact pair is between 0.8 and 4 m^2/N [2]. In this section, single-factor wear simulation studies on coal's Poisson's ratio, shear modulus, density and hardness are conducted from nine levels to obtain the wear law of the middle trough under various physical properties. The overall wear test design is shown in Table 4.6.

The abrasive wear model shown in Fig. 4.26 was used instead of the scraper conveyor model to shorten the simulation time and improve efficiency.

In this section, the wear simulation study of coal particle with the magnitude of 5 mm under different intrinsic parameters is studied. The total mass of coal bulk in the trough was 1 kg, the center of the scraper sample was 110 mm away from the center of the trough, the angular velocity of the ledge was 6.3636 rad/s and the simulation time was set to 5 s. Under this test condition, the speed of the center of the scraper sample was 0.7 m/s, and the trough rotated counterclockwise two times at a constant speed.

In theory, the wear of the middle plate sample is not completely uniform. In order to reduce the impact of the randomness of abrasive wear, the EDEM post-processing function is used to derive the average wear depth of the six sector samples.

Table 4.6 Wear simulation test design

Properties	Value change level								
Poisson's ratio	0.26	0.28	0.30	0.32	0.34	0.36	0.38	0.40	0.42
Shear modulus/10^8 Pa	1	2	3	4	5	6	7	8	9
Density/kg/m^3	1100	1200	1300	1400	1500	1600	1700	1800	1900
Wear constant/10^{-12} m^2/N	0.8	1.2	1.6	2.0	2.4	2.8	3.2	3.6	4

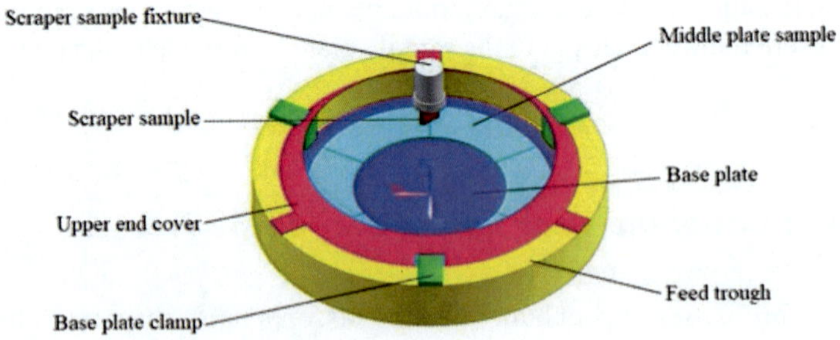

Fig. 4.26 Model of abrasive wear machine

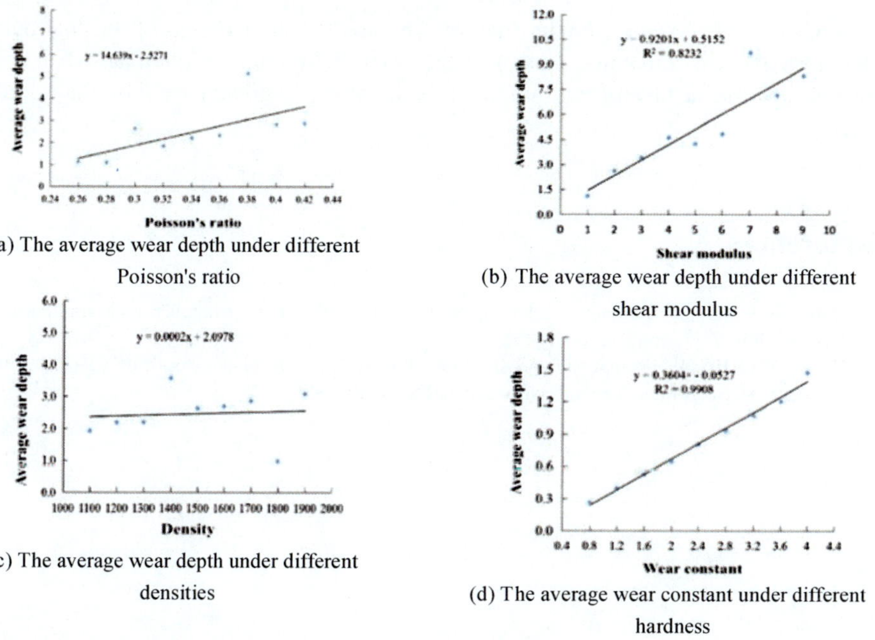

(a) The average wear depth under different Poisson's ratio

(b) The average wear depth under different shear modulus

(c) The average wear depth under different densities

(d) The average wear constant under different hardness

Fig. 4.27 Average wear depth under different physical conditions

According to the values in Table 4.6 at nine levels, the average wear depth chart under different physical properties is obtained as shown in Fig. 4.27.

It can be seen from Fig. 4.27a and b that Poisson's ratio and shear modulus are positively correlated with wear. This is because the contact of coal bulk in the simulation is based on the soft ball model. The larger the Poisson's ratio and shear modulus are, the greater the normal overlap is when coal bulk and geometry contact, which increases the normal contact force of the coal bulk when it comes into contact with the medium sample, and then results in more severe wear.

It can be seen from Fig. 4.27c that changes in density cause significant fluctuations in wear changes. As far as the overall data results are concerned, the greater the density, the greater the amount of wear. When studying the effect of density on wear, the shape and size of individual coal bulk will not change, but the quality of individual particles will increase due to the increase in density. At the same speed, the larger the mass of the coal bulk, the greater the kinetic energy of the single particles. During the process of entering the wedges and forming the three-body wear process, a larger impact load is generated on the middle plate sample. The normal load between the coal bulk and the middle plate sample increases, causing the middle plate sample to wear more seriously.

It can be seen from Fig. 4.27d that the amount of wear increases with the increase of the wear constant, and the wear constant has a strong linear relationship with the amount of wear. The hardness of coal changes the relationship between the friction pair of the coal and steel contact surface. The greater the hardness of the coal bulk, the more likely it is to cause loss of surface material on the middle plate when coal bulk scrape the surface of the middle plate, resulting in greater wear on the middle plate.

References

1. Valentin L (2011) Contact mechanics and friction physical principles and applications. Dissertation. Tsinghua University Press
2. Zhang Y (2016) Wear analysis and structure improvement on bucket tooth of wk-75 mining excavator. Dissertation. Taiyuan University of technology

Chapter 5
Application of DEM on the Design of Wear-Resistant Bionic Chute of Scraper Conveyor

Check for updates

The structure of organism itself is not only related to its heredity but also closely related to its surrounding environment [1]. After hundreds of millions of years of natural selection, the surface structure and physiological mechanism of organisms have developed harmoniously with the natural environment, and gradually formed a special function that can adapt to the harsh environment. Bionics has been widely used in engineering. Many studies have shown that the pit can improve the dynamic lubrication performance of the friction pair, thus reducing the friction and wear of the friction pair and improving the life of the friction pair, but the research on the pit surface under the dry friction condition is still lacking. Some research results show that under the dry friction condition with larger impact, the pit surface has a certain effect on improving the wear resistance of friction parts [2–4], and the arrangement of pits can also improve the stress distribution of friction parts [5]. Because the scraper conveyor is often impacted and cut by coal, the stress distribution on the surface of the middle plate is uneven, and the friction of the scraper and coal on the surface of the middle plate makes to the wear of middle plate very serious.

5.1 Geometric Structure Characteristics of Wear-Resistant Organism Surface

5.1.1 Selection of Biological Geometric Structure Surface and Bionic Model

There are thousands of surface forms of biological geometric structure, and different living environment and function will have different geometric structure. There is

X. Wang et al., *Engineering Applications of Discrete Element Method*, Engineering Applications of Computational Methods 4,
https://doi.org/10.1007/978-981-15-7977-6_5

Table 5.1 Wear-resistant mechanism of biological structure

Wear types	Biological surface	Biologic form	Wear resistance mechanism
Abrasive wear	Dung beetle head	Circle surface-irregularities	Reduce soil resistance
	Land turtle back	Pit	Prevent stress transmission
	Pangolin scales	Corrugated ribs	Generate guidance and scrolling effects
	Shark back scales	Parallelogram shape (or diamond shape)	Provide friction anisotropy
Erosion loss	Scales on the back of the shark	Ridge	Particle flow decomposition
	Scales on the back of the shark	Peak	Reduce contact friction

serious impact abrasive wear in the coal-conveying process of scraper conveyor, so it is more suitable for this subject to use the organism with anti-wear function as the bionic model. The structures of wear-resistant organisms are generally convex hull (such as dung beetle), pit (such as dung beetle), stripe (fan shell, pangolin scale surface) scale (pangolin) and so on. The wear-resistant principle of different structures is shown in Table 5.1. In this study, pit shaped (dung beetle) is selected as the bionic model.

5.1.2 *Pit-Shaped Biological Non-smooth Geometric Structure Surface*

5.1.2.1 Feature Extraction of Non-smooth Elements on the Backplane of Dung Beetle

(1) Focused morphology recovery technology

The focus morphology recovery technology is based on the relationship among camera focal length, image distance and object distance. In order to achieve the purpose of transforming the two-dimensional sequence images to the three-dimensional shapes, the camera moves equidistant along the direction perpendicular to the measured object surface, and collects all the sequence images of the measured object surface.

The imaging principle of the classic optical system is shown in Fig. 5.1. The relationship among object distance u, focal length f and image distance v is as follows:

$$\frac{1}{\mu} + \frac{1}{\upsilon} = \frac{1}{f} \tag{5.1}$$

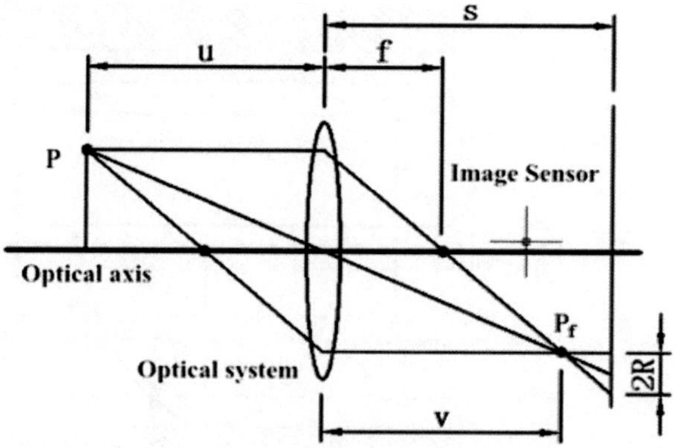

Fig. 5.1 Optical imaging principle

When the focus plane coincides with the image sensor, a clear point P_f can be formed, and when it does not coincide, a blur point with a radius R can be formed. According to this principle, the focus shape restoration process is as follows. Moreover, the flowchart is shown in Fig. 5.2.

(a) Turn on the metallographic microscope, adjust the light source, adjust the camera height to a slightly higher position above the measured object, move the distance between the camera and the measured object equidistantly and ensure that the K frame sequence image covers the entire solid surface.
(b) Remove the non-correlated area of the measured object image and filter the collected image.

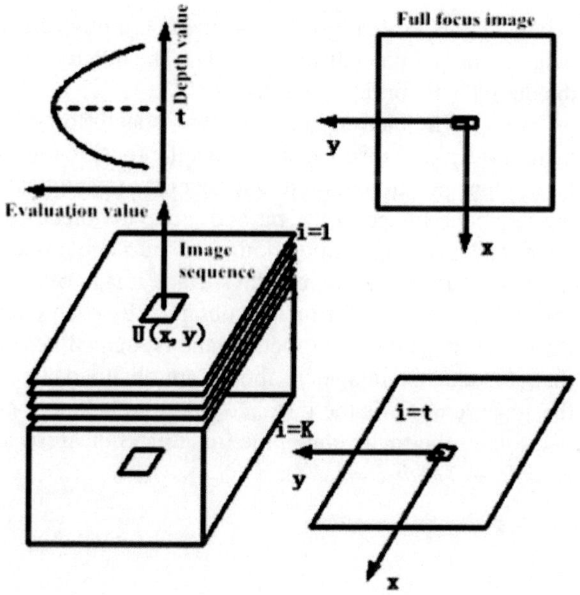

Fig. 5.2 Flowchart of focusing morphology recovery

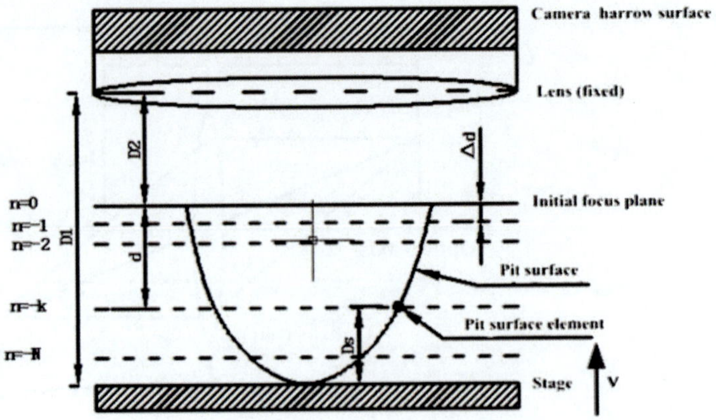

Fig. 5.3 Principle of pit depth measurement

(c) Calculate the focus evaluation value $U(x, y)$ of each pixel (x, y), and obtain the image number corresponding to the maximum point on the focus evaluation curve, $f^*(x, y) = U_{\max}(x, y)$, then obtain the depth matrix $D(x, y, k^*)$ corresponding to the sequence image, and finally the depth figure is processed continuously.

(d) After the initial depth map of the measured object is obtained, the background area outside the outline of the measured object is hollowed out, and finally the three-dimensional shape restoration figure of the measured object is obtained.

(2) Depth measurement method

When a focus evaluation value is the maximum (the point is the clearest), the depth value of the point needs to be known. The depth measurement schematic diagram of the dung beetle pit is shown in Fig. 5.3.

The specific measurement process is as follows: place the dung beetle under the metallurgical microscope and adjust the focal length of the camera so that the initial focus plane position is slightly higher than the highest position of the pit; then collect the sequence image of the pit and move the metallographic microscope stage. The upper and lower adjustment knobs of the stage move a fixed distance Δd each time. Each time the stage moves, the surface image of the pit is to be photographed and saved. The sequence of images obtained by each photograph must be fixed. Repeat the above steps until the focus plane is located below the lowest part of the pit, so that all sequence images of the pit are obtained. D_1 is the distance from the lens to the stage, and D_2 is the distance from the lens to the initial focus plane, when $n = -k$, S is in the focus plane, the imaging is clearest, and the measured depth of S is $D_s = D_1 - D_2 - d (d = k\Delta d)$.

5.1.2.2 Morphology Recovery and Analysis of Non-smooth Elements

(1) MATLAB image processing

Due to the effects of equipment vibration, noise and uneven light source intensity during the actual measurement, the quality of the picture is degraded. Therefore, in order to obtain a high-quality image, the image needs to be cropped and zoomed, filtered to remove noise and performed image segmentation. Through MATLAB software, the image processing program is used to process the sequence image of the pit.

(2) Non-smooth structure and analysis

The specimen of dung beetle was observed under metallographic microscope, as shown in Fig. 5.4a. The surface non-smooth structure of dung beetle magnified 50 times under the microscope is shown in Fig. 5.4b. It can be seen from the observation that its distribution is non-uniform, and the pit mouth is round or oval. In order to further understand the characteristics of non-smooth cells, some of the deep pit cells are selected as the research object by using focused topography restoration technology, and their topography is restored. As shown in Fig. 5.4c and d, the cross-section of the pit is cone-shaped, and the mouth of the pit is basically arc-shaped, so the pit is a cone-shaped round mouth structure. In Fig. 5.4d, the horizontal coordinate

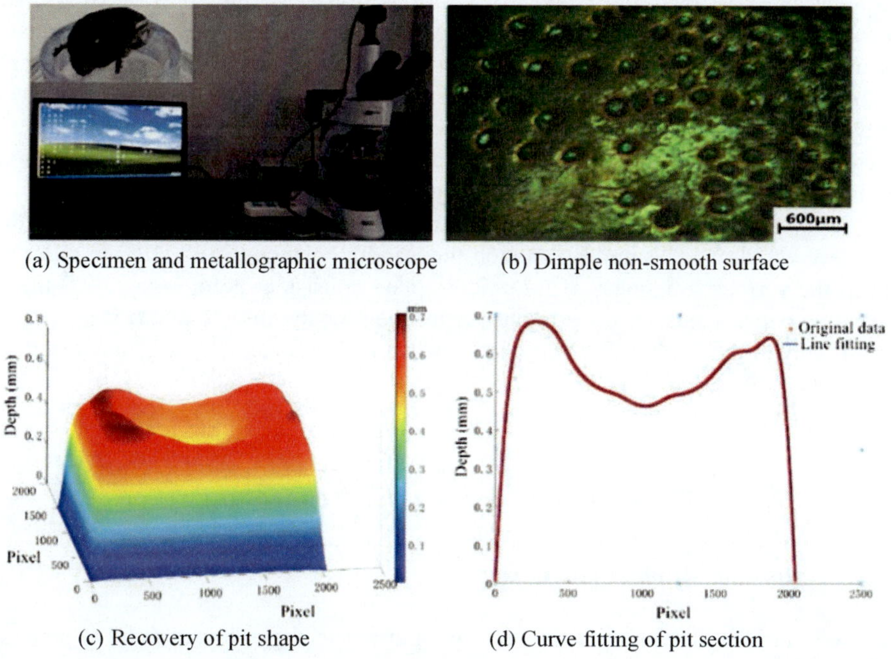

(a) Specimen and metallographic microscope (b) Dimple non-smooth surface

(c) Recovery of pit shape (d) Curve fitting of pit section

Fig. 5.4 Characteristics of dung beetle

Table 5.2 Test conditions

Experimental parameters	Selected value
Coal bulk diameter/mm	6–8
Normal load/N	20
Scraper chain speed/m/s	0.65
Test duration/s	6600

2500-phase element corresponds to 0.25 mm, and it can be seen from the figure that the ratio of pit unit diameter to depth (depth/diameter ratio) is close to 1.

5.2 Optimization of Wear-Resistant Structure of Pit-Type Non-smooth Surface

5.2.1 Test Conditions and Methods

The working condition and environment of scraper conveyor in underground operation is very complex, which is the joint action of multiple working conditions (particle size of bulk material, scraper loading, scraper chain speed, moisture content, gangue content and coal type, etc.), and has a serious impact on the abrasion of scraper conveyor [6]. In order to simulate some working conditions in the mine, facilitate sample cleaning and reduce the random error of the test. The coal material without water and gangue was selected for the test. The allowable parameters of the test environment and equipment were as follows: bulk material size: 1–10 mm, load: less than 35 N, scraper chain speed (feed tank speed): 0.4–0.9 m/s. The working condition values selected in this test are shown in Table 5.2.

The wear of the middle plate is the difference between the mass before and after the test, which is measured by one ten-thousandth balance (FA3204B). In order to keep the surface roughness of the middle plate consistent before wear, 600-mesh abrasive paper was used for grinding. After the test, the middle plates were wiped with anhydrous alcohol and the coal powder in the pit unit was cleaned with a high-pressure air gun.

5.2.2 Bionic Middle Plate Design and Preparation

5.2.2.1 Bionic Middle Plate Design

A section of the pit element is in the form of arc cone angle. It is difficult to prepare the arc pit element on the bionic plate. Considering the characteristics of the tool for machining the pit, the conical pit is simplified as a cylindrical pit with the bottom cone angle of 120°. The pit elements of dung beetle are distributed unevenly, but in order to adapt to the wear machine samples, they are arranged in a fan-shaped uniform way. The ratio of depth to diameter is a typical characteristic of pit, which is regarded as a characteristic factor affecting wear. According to the research of grinding roller on the pit surface of Xi et al. [7] and the simulation research of abrasive wear on the pit surface of Liu et al. [8], the pit diameter, the distance between pits in the direction of motion (pitch angle) and perpendicular to the direction of motion (radial distance) of the grinding parts were selected as the factors affecting the wear amount. The depth/diameter ratio, pit diameter, pitch angle and radial distance are, respectively, represented by ρ, D, A and L. Figure 5.5 shows the design sample diagram of bionic board.

Depth/diameter ratio ($\rho = h/d$): The longitudinal section of the pit unit of dung beetle is square, that is, the depth/diameter ratio is close to 1, and six levels were set at the center of 1: 0.3, 0.6, 0.9, 1.2, 1.5 and 1.8.

Pit diameter (D): Determine the diameter range according to the size of the test sample. The thickness of the plate is 3.7 mm, so that the pit is a blind hole, then the

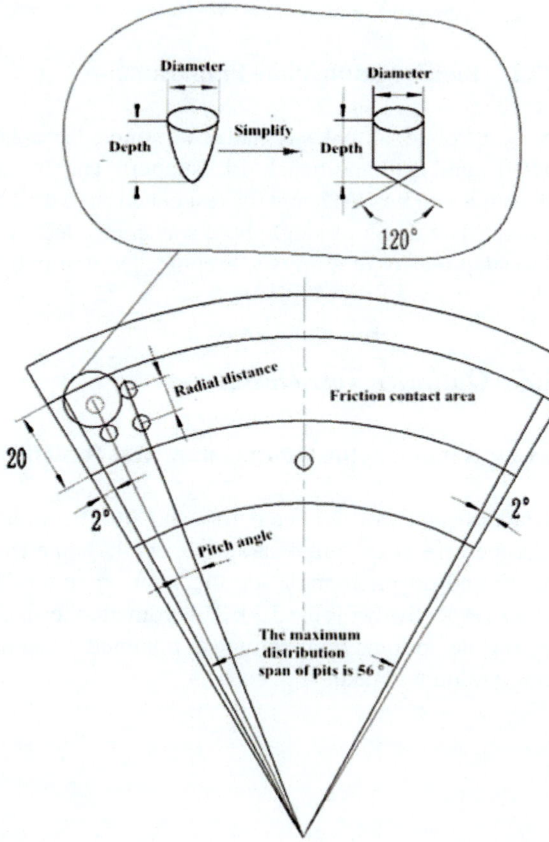

Fig. 5.5 Schematic drawing of pit plate design

diameter meets the following formula:

$$\begin{cases} 1.8D \leq 3.7\text{mm} \\ D \leq 2.06\ \text{mm} \end{cases} \tag{5.2}$$

Take six levels uniformly: 0.4, 0.7, 1.0, 1.3, 1.6 and 1.9.

Pitch angle (A): The span angle of the middle plate is 60°. In order to reduce the test error, the pits are not arranged at 2° before and after the middle plate, so the maximum span of the actual pit arrangement area is 56°. At the same time, in order to include the limit range, the maximum pitch angle is 14° when five rows of pits are arranged in the moving direction of the upper specimen. Take six levels evenly: 2.5°, 4.5°, 6.5°, 8.5°, 10.5° and 12.5°.

Radial distance (L): The width of the grinding part of the upper sample is 20 mm, so the actual action width of the upper sample and the lower sample is about 20 mm, that is, the width of the friction contact area is about 20 mm. At the same time, ensure that there are at least two pits in the radial direction, then evenly take six levels: 3, 4.5, 6, 7.5, 9 and 10.5.

5.2.2.2 Bionic Middle Plate Preparation

The middle plate sample was made by a simple three-axis milling machine composed of J10-1 hand drill and bench drill support. The layout and position of all pit units shall be marked by CAD, and the test bench shall drill holes according to the marked positions. The depth of single hole was controlled by ST5-1B pointer depth meter, and the depth error of all pits was controlled within 0.03 mm.

5.2.3 Optimize Test Design

5.2.3.1 Single-Factor Optimization Test Design

Taking the quality difference before and after the middle plate wear as the response value, the effects of four factors, pit depth/diameter ratio, pit diameter, pit radial distance and pit pitch angle, on the wear amount of the bionic middle plate were studied, respectively (Table 5.3). According to the single-factor optimization results, a reasonable optimization region was obtained. Each factor was tested twice, and the average value was taken.

Table 5.3 Single-factor optimization design level

Factor	Levels
Pit depth to diameter ratio (ρ)	0.3, 0.6, 0.9, 1.2, 1.5, 1.8
Pit diameter (D)/mm	0.4, 0.7, 1.0, 1.3, 1.6, 1.9
Pit pitch angle (A)/°	2.5, 4.5, 6.5, 8.5, 10.5, 12.5
Radial distance of pit (L)/mm	3, 4.5, 6, 7.5, 9, 10.5

Table 5.4 Response surface optimization design level

Factor	Levels		
	−1	0	1
Pit depth to diameter ratio (ρ)	1.1	1.3	1.5
Pit diameter (D)/ mm	0.5	0.7	0.9
Pit pitch angle (A)/°	5	6.5	8
Radial distance of pit (L)/ mm	3.75	4.5	5.25

5.2.3.2 Response Surface Method to Optimize Test Design

The result of single-factor optimization has deviation, response surface method can further improve the accuracy of optimization and the result data can be analyzed comprehensively. For example, the significance of factors can be analyzed, and the factor interaction law of response surface can be analyzed, so the response surface method test design is carried out. According to the design principle of Box-Behnken, response surface method was used to design the experiment. And each factor takes three levels of high (1), medium (0) and low (−1). The design center uses three central points for error estimation. The test level design is shown in Table 5.4.

5.2.4 Experimental Results and Analysis

5.2.4.1 Single-Factor Test Results and Analysis

The single-factor test investigated the influence of the depth to diameter ratio, diameter, pitch angle and radial distance of the pit on the wear of the bionic plate.

From Fig. 5.6a, it can be seen that the pit diameter increases within the range of single factor, and the wear amount of bionic plate decreases gradually. When it reaches 1.3, the change is gentle. Therefore, 1.3 was taken as the 0 value of response surface method test. It can be seen from Fig. 5.6b that the pit diameter increases, and the wear amount is relatively low when the pit diameter is 0.7 and 1.6 mm, respectively, but when the pit diameter is 1.6 mm, the crushing rate of the sample to the coal bulk is large, and because the pit is not filled by the coal powder in the early stage of the test, the impact of the pit with the larger diameter

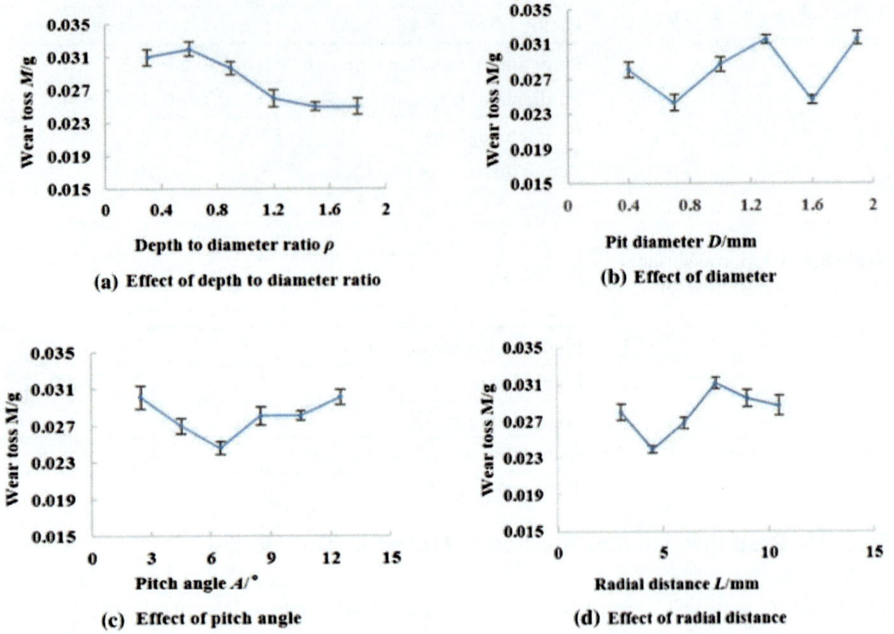

Fig. 5.6 Effects of various factors on wear

of 1.6 mm and the upper sample is more intense. Therefore, considering comprehensively, the pit diameter response surface method optimization test 0 value was taken as 0.7 mm. It can be seen from Fig. 5.6.c and d that in the range shown in the figure, the amount of wear decreases first and then rises with the pitch angle and radial distance. The minimum amount of wear is at a pitch angle of 6.5° and a radial distance of 4.5 mm. Therefore, the pitch angle and radial distance were, respectively, taken as 6.5° and 4.5 mm as the response surface method 0 value. The absolute value of the difference between the two tests at the same level represents the data fluctuation amplitude. The smaller the amplitude, the higher the accuracy. The maximum differences of the depth/diameter ratio, diameter, pitch angle and radial distance are 0.0020, 0.0018, 0.0024 and 0.0022, respectively. The minimum differences are 0.0010, 0.0008, 0.0012 and 0.0008, respectively.

5.2.4.2 Response Surface Test Results and Analysis

(1) Box-Behnken test and regression model

The experiments and results designed according to the Box-Behnken principle are shown in Table 5.5. Based on the experimental results, a second-order regression model of the wear amount of middle plate and four factors was established using

Table 5.5 Box-Behnken test design and results

Test number	Depth to diameter ratio (ρ)	Diameter (D)/mm	Pitch angle (A)/°	Radial distance (L)/mm	Wear rate (M)/g
1	0.00	−1.00(0.5)	0.00	1.00	0.0264
2	1.00	0.00(0.7)	0.00	−1.00(3.75)	0.0284
3	0.00	0.00	−1.00(5)	−1.00	0.0300
4	−1.00(1.1)	0.00	−1.00	0.00(4.5)	0.0279
5	0.00(1.3)	−1.00	−1.00	0.00	0.0300
6	0.00	0.00	0.00(6.5)	0.00	0.0250
7	0.00	−1.00	0.00	−1.00	0.0324
8	−1.00	0.00	0.00	−1.00	0.0326
9	−1.00	1.00(0.9)	0.00	0.00	0.0276
10	1.00(1.5)	1.00	0.00	0.00	0.0263
11	0.00	1.00	1.00(8)	0.00	0.0276
12	0.00	1.00	0.00	−1.00	0.0265
13	−1.00	−1.00	0.00	0.00	0.0291
14	−1.00	0.00	1.00	0.00	0.0285
15	−1.00	0.00	0.00	1.00(5.25)	0.0268
16	0.00	0.00	−1.00	1.00	0.0257
17	0.00	1.00	−1.00	0.00	0.0249
18	0.00	0.00	0.00	0.00	0.0245
19	0.00	0.00	1.00	1.00	0.0266
20	1.00	−1.00	0.00	0.00	0.0250
21	0.00	−1.00	1.00	0.00	0.0258
22	0.00	0.00	1.00	−1.00	0.0275
23	0.00	1.00	0.00	1.00	0.0278
24	1.00	0.00	1.00	0.00	0.0240
25	0.00	0.00	0.00	0.00	0.0239
26	1.00	0.00	0.00	1.00	0.0252
27	1.00	0.00	−1.00	0.00	0.0254

Design expert software.

$$\begin{aligned} M = 0.32877 - 0.10542\rho - 0.171D - 0.010218A \\ -0.058294L + 0.0175\rho D - 0.00166667\rho A \\ +0.00433333\rho L + 0.00575DA \\ +0.012167DL + 0.000755556AL \\ +0.029583\rho^2 + 0.037708D^2 \end{aligned}$$

Table 5.6 ANOVA of quadratic polynomial model of Box-Behnken test

Source of variance	Sum of square	Degree of freedom	Mean square	F value	P value
Model	1.292E-004	14	9.232E-006	15.41	<0.0001
A-ρ	2.760E-005	1	2.760E-005	46.07	<0.0001
B-D	5.333E-006	1	5.333E-006	8.90	0.0114
C-A	1.268E-006	1	1.268E-006	2.12	0.1714
D-L	2.977E-005	1	2.977E-005	49.69	<0.0001
AB	1.960E-006	1	1.960E-006	3.27	0.0956
AC	1.000E-006	1	1.000E-006	1.67	0.2207
AD	1.690E-006	1	1.690E-006	2.82	0.1189
BC	1.190E-005	1	1.190E-005	19.87	0.0008
BD	1.332E-005	1	1.332E-005	22.24	0.0005
CD	2.890E-006	1	2.890E-006	4.82	0.0485
A^2	7.468E-006	1	7.468E-006	12.47	0.0041
B^2	1.213E-005	1	1.213E-005	20.25	0.0007
C^2	3.593E-006	1	3.593E-006	6.00	0.0306
D^2	2.873E-005	1	2.873E-005	47.95	<0.0001
Residual	7.189E-006	12	5.991E-007		
Mismatch	6.469E-006	10	6.469E-007	1.80	0.4100
Pure error	7.200E-007	2	3.600E-007		
Sum	1.364E-004	26			

$R^2 = 0.9473$, $CV = 2.86\%$, Adeq Precision = 14.271

$$+0.000364815\text{A}^2+0.00412593\text{L}^2 \tag{5.3}$$

Table 5.6 shows the variance analysis of the regression model, from which it can be seen that $P < 0.01$, indicating that when the model describes the relationship between factors and response values, the relationship between dependent variables and all independent variables is very significant. The mismatch term $P = 0.4100 > 0.05$, which shows that the equation fits well. From Table 5.6, it can be seen that pitch angle (A) has no significant effect on wear. The depth/diameter ratio (ρ), diameter (D) and radial distance (L) are significant, and the depth/diameter ratio and radial distance P value are less than 0.01, which is extremely significant. The order of significance of each factor is $L > \rho > D > A$. Generally, the higher the coefficient of variation (CV), the lower the reliability of the test. The coefficient of variation in this test is 2.86% less than 10%, indicating that the prediction model of the amount of wear and various factors has high reliability. The determination coefficient $R^2 = 0.9473$, which is close to 1, indicates that the fitting equation has high reliability. The precision is 14.271 greater than 4, indicating that the model has good accuracy. Based on the analysis results, this polynomial regression model can be used to predict and optimize the amount of bionic plate wear.

(2) Analysis of the interaction effect of regression model

The three-dimensional response surface graph can express the functional relationship between the response value and each level of the test factors, and the optimal parameters in the test design can be obtained from an intuitive perspective. The response surface and contour figure of the test are shown in Fig. 5.7. It can be seen from the figure that the response surfaces of interaction terms have minimum values, and the response surfaces are all reverse parabola. Figure 5.7b shows that the interaction between radial distance and diameter has the most significant effect on the wear of the middle plate. When the radial distance and diameter are greater than zero, the wear increases with the increase of the two parameters. When the radial distance and diameter are less than zero, the wear increases with the decrease of the two parameters, but the wear increases with the decrease of the two parameters more steeply than the increase of the two parameters. The interaction between pitch angle and diameter, radial distance and pitch angle, and radial distance and diameter have similar laws on the wear of the middle plate. The change of pitch angle and radial distance reflects the distance between pits. If the spacing is too large, the stress concentration on the surface of the middle plate will be obvious [9], and the wear will be intensified. If the distance is too small, the impact between the grinding parts will be intensified, and the wear will also increase, so the reverse parabola will appear.

5.2.5 *Verification and Comparative Test of Bionic Middle Plate with Optimal Structure*

Through response surface optimization, it is found that when the depth diameter ratio, diameter, pitch angle and radial distance are 1.41, 0.69 mm, 6.55° and 4.66 mm, respectively, the wear of the middle plate is the smallest.

In order to verify the reliability of the model, the working condition parameters in Table 5.2 were used. Due to the consideration of the actual processing conditions, it is necessary to adjust the optimized diameter results from 0.69 to 0.7 mm, and keep the others as the optimized values. The optimized bionic board experiment was repeated three times, and the test results and theoretical values are shown in Table 5.7. The relative error is 3.2%, and the measured value is close to the test value.

In order to test the wear degree of the optimized bionic middle plate and the smooth middle plate, the abrasive wear test of the smooth middle plate was carried out and compared with the wear test results of the optimized middle plate. The smooth middle plate test was repeated three times and the average value was taken. The test results are shown in Table 5.7. The wear of bionic middle plate was reduced by 12.6% compared with the smooth middle plate.

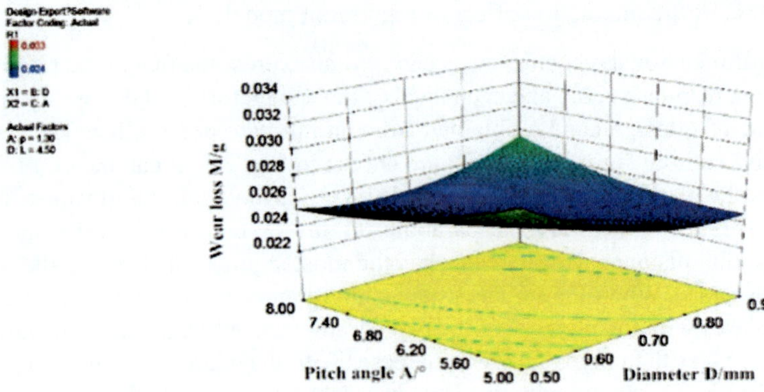

(a) Pitch angle and diameter interaction

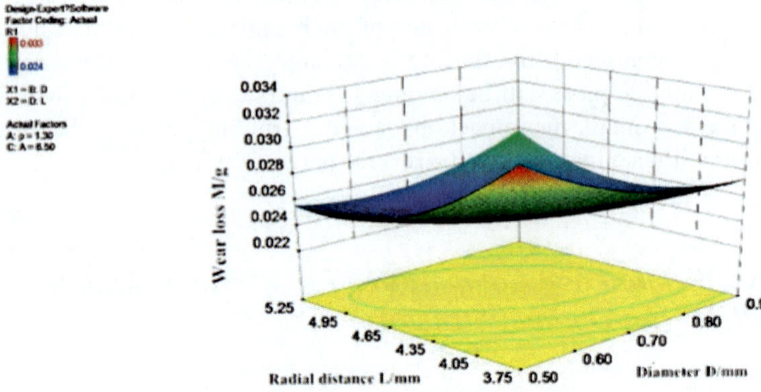

(b) Radial distance and diameter interaction

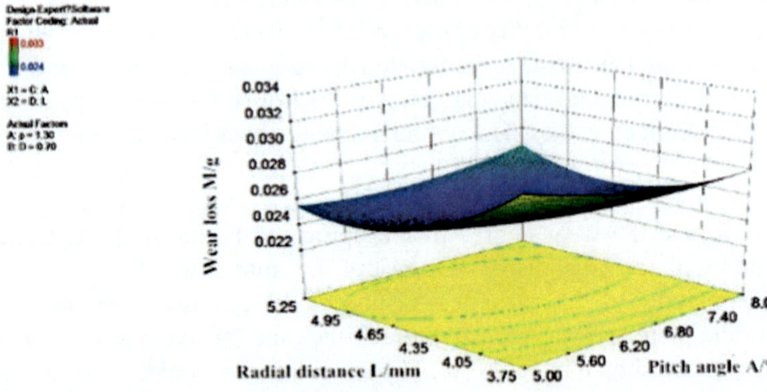

(c) Radial distance and pitch angle interaction

Fig. 5.7 Factor interaction diagram

Table 5.7 Verification and comparison

Middle plate types	Predictive value/g	Experimental value/g		Mean value/g	
Optimize bionic plate	0.0239	0.0243	0.0253	0.0246	0.0247
Smooth plate	–	0.0282	0.0279	0.0288	0.0283

5.3 EDEM Simulation Analysis of the Wear-Resistant Mechanism of Bionic Middle Plate with Optimal Structure

Many researches show that the bionic non-smooth surface has a good effect on improving the wear resistance of the friction pair surface. This paper studies the wear resistance effect of the bionic pit and striped non-smooth surface on the middle plate of scraper conveyor, reveals the wear-resistant mechanism and wear mechanism of the bionic non-smooth surface, which is of great significance to further optimize the bionic wear resistance design. Different wear-resistant mechanism and wear mechanism will produce different micro-wear morphology on the friction pair surface. Therefore, this paper studies the micro-wear morphology of the optimal wear bionic middle plate surface and reveals the wear-resistant mechanism by comparing with the smooth middle plate surface. The middle plate of scraper conveyor is subject to friction and wear of scraper or scraper chain. At the same time, the middle plate and scraper or scraper chain and coal bulk are also subject to serious three-body wear. Therefore, the middle plate and scraper, scraper chain and coal bulk have a complex mechanical relationship. The mechanical properties of the scraper and the middle plate in the wear process of the middle plate are studied by DEM-MBD, and it can provide great convenience for the study of wear resistance mechanism.

5.3.1 Research on Wear Mechanism Based on Wear Morphology

The wear effect of the pit middle plate before and after the test is shown in Fig. 5.8.

Figure 5.9 shows the untreated pit morphology of bionic middle plate after the test. It can be seen from Fig. 5.9 that the coal bulk material is densely filled in the pit after crushing, and the newly filled pit presents two forms: in Fig. 5.9a, the coal bulk or coal powder form a new pit with a shallow depth and a circular arc at the bottom of the pit, and the rolling of the coal bulk stored in the pit improves the relative movement state of the upper and lower samples; in Fig. 5.9b, coal bulk or pulverized coal fill the pit completely. The hardness of the dense coal powder in the pit is lower than that of the surrounding metal, which plays a buffer role for the impact of abrasive. The deformation of the pit can improve the stress condition of

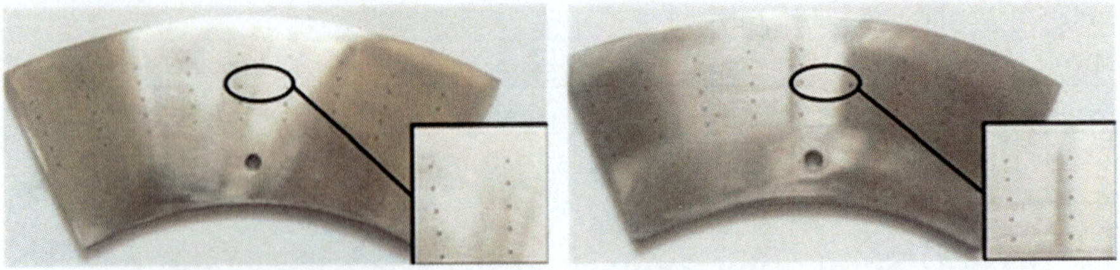

(a) Before the test (b) After the test

Fig. 5.8 Pit plate after test

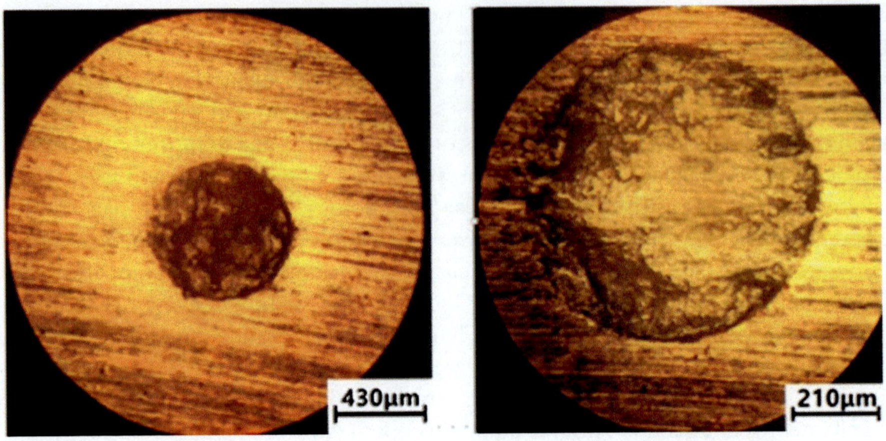

(a) Dense pit state (b) Dense filling state

Fig. 5.9 Appearance of the pit after wear

the middle plate. At the same time, the dense coal powder can prevent the edge of the pit from being excessively deformed, which in turn leads to flaking.

Figure 5.10 shows the wear morphology of the smooth and optimized bionic middle plate after the test, and the arrow indicates the moving direction of the middle plate. It can be seen from Fig. 5.10a that the furrow on the surface of the smooth middle plate is more obvious and the depth and width are different, that is, serious abrasive wear has occurred. The main reason is that the small coal bulk are sandwiched between the upper sample and the middle plate. When the coal bulk form a certain angle with the middle plate, the sharp coal bulk edges and corners cut the surface of the middle plate to form a furrow. It can also be seen that the smooth surface of the middle plate presents a long and thin or water-drop-type metal adhesive block, because when the small coal bulk micro cutting the metal matrix, the metal matrix will be squeezed to the two sides of the coal bulk. A large number of coal bulk move and extrude in a circular way, which makes the materials on the surface of the middle plate peel off and form debris. When the metal debris is in the high pressure, high

(a) Wear profile of smooth middle plate (b) Radial wear profile of bionic middle plate pits

(c) Wear appearance at the exit pit (d) Wear appearance when entering the pit

Fig. 5.10 Micro-cosmic wear morphologies of smooth middle plate and bionic middle plate

speed and high temperature state, the surface of the middle plate is sticky touch and displace.

Figure 5.10c and d shows the wear morphology of the entry and exit pits, respectively, and the furrows with different depth and width can be seen. Compared with the smooth plate, the furrow is uniform and the width is smaller, and the wear degree of abrasive particles is lighter, and there are more impact pits formed due to the change of coal bulk attitude at the exit pit. Because the coal bulk on the surface of the middle plate meet the pit in the continuous cutting, the state of continuous cutting is destroyed and rolling occurs on the surface of pit, so more pitting is formed at the exit of the pit. The interruption of continuous cutting can improve the wear resistance.

Figure 5.10b shows the wear morphology among the radial pits of the bionic plate. It can be seen from Fig. 5.10b, c and d that the adhesion block on the surface is significantly less than that on the smooth plate. According to the theory of exfoliation wear [10], when the hard contact peak point and each point on the soft surface are subjected to multiple cyclic loads, the surface of the metal soft surface will produce shear plastic deformation and accumulate continuously, which will form dislocation and stress concentration phenomenon on the surface or subsurface. The distribution of pits can greatly improve the surface stress distribution of the middle plate, and

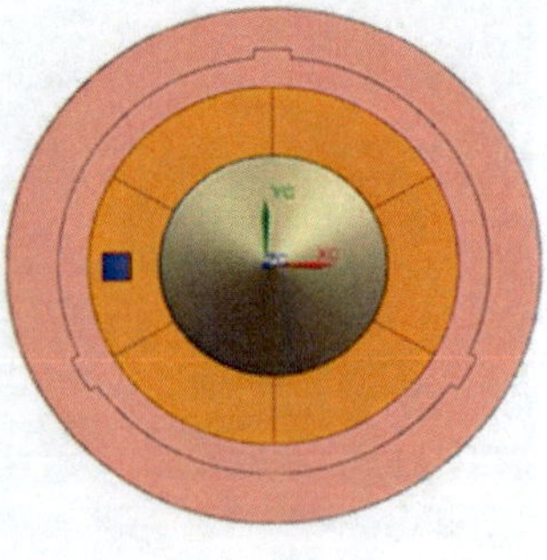

Fig. 5.11 Pit plate arrangement

then improve the contact peak stress state of the metal surface, which can not only reduce the fatigue wear but also reduce the peeling off of wear debris, reduce the occurrence of adhesion phenomenon and improve the wear resistance. It can also be seen from Fig. 5.10b that the furrow among the radial pits is more uniform and less wide than that of the smooth plate, and the wear degree of the abrasive particles is weaker.

5.3.2 Research on Wear Resistance Mechanism of EDEM-RecurDyn Coupled Simulation

The simulation duration was 1.32 s (particle generation time is 0.5 s; feed tank rotation time is 0.82 s). The pit-shaped bionic middle plate and the smooth plate were arranged in the manner as shown in Fig. 5.11. The bionic plate and the smooth plate were, respectively, arranged in two pieces. The periods of wear of the middle plate, upper sample and coal bulk were: bionic plate 0.58 ~ 0.76 s, 0.94 ~ 1.12 s, smooth plate 0.76–0.94 s, 1.12–1.3 s, so only analyze the simulation data between 0.58 and 1.3 s. The force of the upper sample is shown in Fig. 5.12a, and the middle

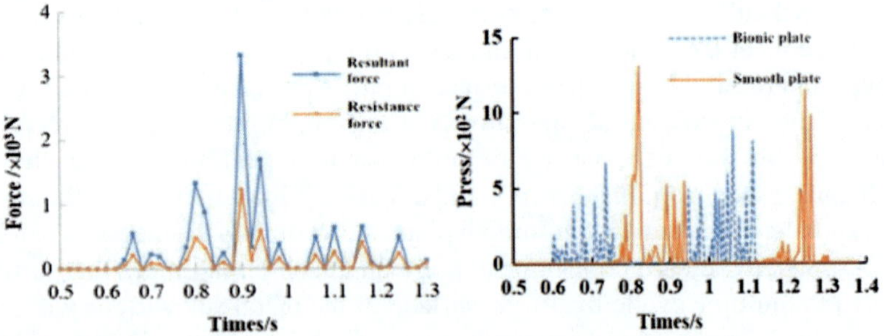

(a) Force applied to the sample (b) Middle plate particle pressure

Fig. 5.12 Result of simulation

plate is subjected to the particle pressure as shown in Fig. 5.12b. The particle size of the coal bulk was 6–8 mm, the load of the upper sample was 20 N and the rotating speed of the feed tank was 5.91 r/s (i.e., the rotating linear speed was 6.5 m/s).

In the process of long-distance transportation of coal by scraper conveyor, the sudden change of torque or chain break of driving device is often caused by the stagnation of coal in the vicinity of scraper. The particle pressure on the surface of the middle plate plays a key role in the wear degree of the middle plate, so it is necessary to study the resistance of the upper sample (scraper) and the particle pressure on the surface of the middle plate.

It can be seen from Fig. 5.12a that the resistance of the upper sample is a process of constant fluctuation, with a large amplitude, and the law of resultant force and resistance is similar. The maximum resistance of the sample on the smooth plate surface is 1243.5 N, while that of the bionic plate is 602.7 N. The maximum resistance of the sample on the surface of bionic middle plate is smaller than that on the smooth plate. It can be seen from Fig. 5.12b that the normal pressure value (hereinafter referred to as pressure) of particles on the surface of the middle plate fluctuates greatly, the maximum pressure of the smooth middle plate is 1310.6 N and the maximum pressure value of bionic middle plate surface is 1072.5 N, which is smaller than the maximum pressure value of the smooth plate surface. The above phenomenon is due to the wedge-shaped structure of the upper sample and the formation of a certain angle with the middle plate, which is easy to be stuck with the coal. When the coal bulk is between the upper sample and the middle plate, the extrusion pressure of the upper sample and the middle plate to the coal bulk becomes larger as the material tank rotates, and the coal bulk will continue to cut the surface of the middle plate. Because of the existence of pits, the bionic plate is more likely to destroy the continuous cutting state of coal bulk, so the peak pressure on the surface of the bionic plate is less than the smooth plate. Figure 5.12b shows that the frequency of pressure change on the bionic plate surface is relatively high, which also shows that the particle attitude change on the bionic plate surface is more frequent.

Figure 5.13 is a discrete element simulation wear cloud chart, on which the red line represents the wear mark (furrow) with the maximum wear depth, and the white

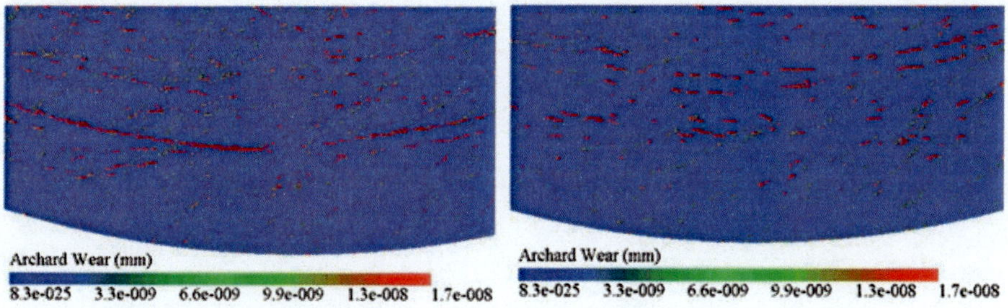

(a) Smooth plate wear cloud (b) Bionic plate wear cloud

Fig. 5.13 Middle plate wear cloud

fine point line represents the pit position. It can be seen that there are continuous and long wear marks on the surface of smooth plate, while the bionic plate wear marks are short and discontinuous, and the wear marks in the pits are all disconnected.

References

1. Kramer K (1965) Bionics. Beijing Science Popularization press, Beijing
2. Yang C (2005) Experimental research on sliding anti-wear ability of the bionic non-smooth sur-face and numerical simulation. Dissertation. Jilin University
3. Yang Z (2006) Research of wear-resistance on roller with biomimetical non-smooth concave sur-face. Dissertation. Jilin University
4. Teng F (2014) Research of roller surface morphology designing and wear resistance and grinding capability of cement roller press. Dissertation. Jilin University
5. Qian Z (2007) Research on stress-strain constitutive relationship for bionic non-smooth wearable surface. Dissertation. Jilin University
6. Shi Z (2017) Wear rule of the middle trough in armoured conveyor under severe condition. J China Coal Soc 42(S2):541–546 (in Chinese)
7. Xi P, Cong X, Teng F, Gou H (2018) Design and test of bionic concave grinding roller for improving wear resistance and crushing. Trans Chin Soc Agric Eng 34(08):55–61 (in Chinese)
8. Liu Y, Wang X, Li B, Wang S, Yang Z (2017) Bionic optimization and wear performance of central groove scraper. Min Res and Development. 37(04):24–27 (in Chinese)
9. Liu Y, Wang X, Li B, Wang S, Wang S, Yang Z (2017) Wear resistance analysis and optimization design of pit-shaped non-smooth surface. Journal of Taiyuan University of Technology. 48(01):73–78 (in Chinese)
10. Wen S, Huang P (2012) Principles of Tribology. Tsinghua University Press, Beijing

Chapter 6
Flowing State Analysis of Coal Bulk in Silos Using DEM

Most of silos are straight cone structure, rectangular pyramid structure and straight cylinder plus hyperbolic discharge hopper structure. But problems such as blockage and poor flow often occur in the specific use of coal silos (or other silos) [1]. For a variety of automatic and manual dredging measures [2, 3], they are either expensive, inconvenient or dangerous; especially the sudden fall during dredging process can cause serious production accidents. The above problems have a great impact on production efficiency and safe operation. In view of the research on the blockage of coal silo, the existing literatures mainly focus on the structure, design, materials, construction, operation, monitoring, prevention and dredging of coal silo. In fact, the blockage or poor circulation of coal silo is not only related to the coal silo itself and operation, but also closely related to the physical properties and flow characteristics of the stored coal. There are many researches on the flow characteristics, hydrodynamic behavior of all kinds of silos and stored bulk materials [4, 5]; especially with the continuous development and maturity of discrete element method, finite element method and other methods, techniques and tools, the numerical simulation method is gradually applied to the study of the flow of bulk material in the silo [6, 7].

6.1 Construction of DEM Model for Silo-Coal Particle Interaction

6.1.1 Coal Particle Model

The generation of bulk particles is very important in the simulation. It is not easy to set the real shape of coal particles in the software, and it is not representative. Therefore,

X. Wang et al., *Engineering Applications of Discrete Element Method*, Engineering Applications of Computational Methods 4,
https://doi.org/10.1007/978-981-15-7977-6_6

Table 6.1 Material parameters of coal particle

Material of particles	Poisson's ratio	Shear modulus	Particle density
Coal	0.25	3×10^7 Pa	1600 kg/m^3

in order to simplify the calculation, circular particles are used to represent the shape of coal particles. In EDEM, the particle size distribution of the generated particles shows a normal distribution, with a mean radius of 100 mm and a variance of 0.05. Other properties of coal particles are shown in Table 6.1.

6.1.2 *Silo Model*

Referring to the cylindrical silo in a coal mine as the original model, the first step was to create the geometric model of the silo in UGNX 7.5 software, in which the geometric parameters set are shown in Table 6.2.

After creating the three-dimensional model, because the EDEM software itself does not have the function of meshing, we should first use other CAE tools to divide the geometric mesh. This section chooses Gambit to deal with (Note: Gambit is a very powerful pre-processing software developed for CFD, which can divide a very high-quality mesh). After the.prt file format the model was converted into.igs format, it was imported into Gambit to divide the mesh, and the mesh density was set to 400. Because the EDEM does not support square mesh elements, it was divided into triangular elements. Imported the generated.msh file into EDEM. Figure 6.1 is a schematic diagram of the geometric parameters of the coal silo model and the final geometric model of the silo shown in EDEM after meshing.

Opened the global panel of the software and set the material of the silo model to reinforced concrete. Its parameters are shown in Table 6.3.

Table 6.2 Geometric parameters of silo

Height of coal silo	Inclination of the hopper	Height of the hopper	Diameter of the discharge port	Diameter of the cylinder
18 m	65°	4.83 m	1.5 m	6 m

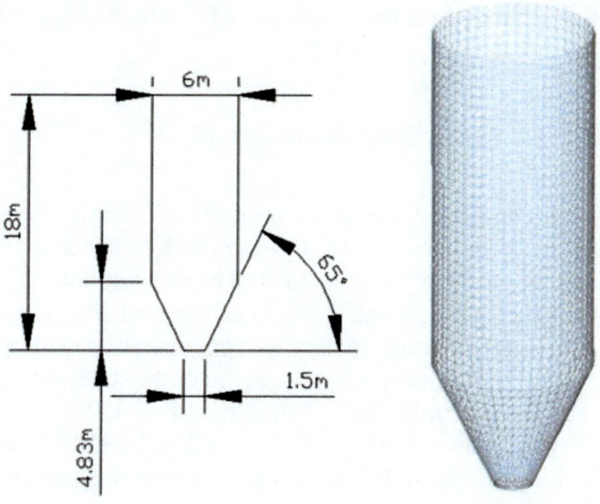

Fig. 6.1 Size and grid style of silo

Table 6.3 Material parameters of silo model

Material	Poisson's ratio	Shear modulus	Density
RC	0.28	1.8×10^{10} Pa	2550 kg/m^3

6.1.3 Contact and Flow Model

In this section, the Hertz-Mindlin (no slip) contact model was selected, and the gravitational acceleration was 9.81 m/s^2. The contact parameters are shown in Table 6.4.

Table 6.4 Contact parameters

Contact parameters	Value
Coefficient of restitution between the coal particles	0.1
Coefficient of static friction between the coal particles	0.45
Coefficient of rolling friction between the coal particles	0.1
Coefficient of restitution between the coal particles and the wall	0.15
Coefficient of static friction between the coal particles and the wall	0.35
Coefficient of rolling friction between the coal particles and the wall	0.1

6.2 Simulation Analysis of Bulk Material Flow State in Silo by DEM

6.2.1 Flow State Simulation

After completing the above process, clicked the Geometry button of the EDEM to set the geometric features of the particle factory. The particle factory should not only ensure that the particles fall completely into the bin body but also ensure that the generated coal particles can fill the coal silo. Therefore, this section selected the shape of the particle factory as a circular sheet with a diameter of 5.6 m, which was placed 17.5 m away from the baffle of the discharge port, as shown in Fig. 6.2.

Next, a circular baffle with a radius of 0.75 m was added to the discharge outlet of the silo model, which was created in UGNX7.5 and imported into EDEM in.igs format after completion, and its position was adjusted under the 'Position' to just block the discharge outlet, and the baffle material was set to steel. The time when the particle factory started to generate the first coal particle was $t = 0$ s, and when $t = 14$ s, the baffle left the unloading port at the speed of 10 m/s, thus opened the unloading port. Before the coal silo was unloaded, all the coal particles need to be filled into the coal silo.

Then clicked the 'Factory' to add a new factory to generate 6000 coal particles dynamically. In addition to gravity, coal particles were also provided with a downward initial speed of 10 m/s. When $t = 12$ s, all the particles were formed. During the falling process, the particles collide with each other, and the kinetic energy of the particles is consumed continuously. Finally, the coal bulk are stored in the coal silo, and after standing for 2 s, the whole system has reached a stable state. Figure 6.3 shows the coal particle generation diagrams at $t = 2$, 4, 6, 9 and 11 s.

Entered EDEM solver module, Rayleigh time step was automatically set to 0.00213 s, and the fixed time step was 25% of Rayleigh time step, which was

Fig. 6.2 Particle factory position in the silo

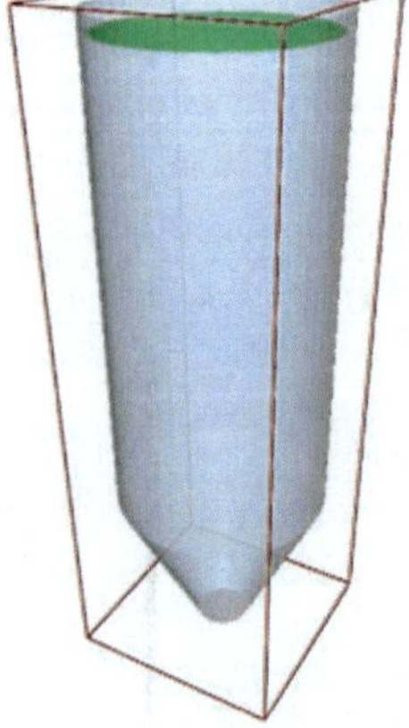

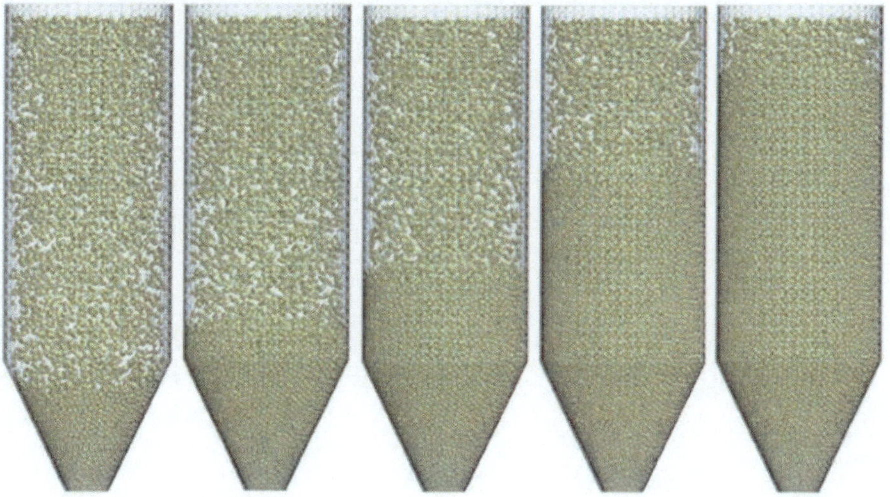

Fig. 6.3 Silo charging process

0.000531449 s. The total simulation time was set to 150 s, the grid size of the simulation area was set to $3R_{max}$, and the simulation area was divided into 42,048 small cubes.

When $t = 14$ s, the baffle at the discharge port left the discharge port, and the coal particles began to flow out of the coal silo under the action of gravity. In order to save the calculation time and the storage space of the hard disk, the particles flowing out were set outside the boundary conditions, so the particles flowing out of the coal silo during the simulation were not in the field of vision. Figure 6.4 shows the unloading process of silo in EDEM.

As shown in Fig. 6.5a, five points with a height of 16 m are marked (other particles are shielded for clarity). Figure 6.5b shows the positions of the five coal particles at 127 s. Figure 6.5c shows the corresponding trajectories of the five coal particles during the process from 14 to 127 s. From the flow situation, it is found that the whole flow occurs at the beginning of unloading, and the falling speed of particles in the same material layer is equal, showing vertical falling basically. When the top surface of the material layer flows to the lower end of the straight cylinder, it becomes central flow, and the particles near the silo wall begin to move toward the center gradually. After flowing to the hopper, the particles on the central axis flow out first, and the particles closer to the hopper wall flow out later.

The above phenomenon shows that the flow speeds of coal particles at the same level in the straight cylinder close to the discharge hopper and inside the discharge hopper are different. The flow speed is the slowest near the silo wall and the fastest in the middle. As the cross-section of the silo becomes smaller and smaller in the discharge process, the coal particles in the straight barrel will show an uneven decrease when flowing to the hopper.

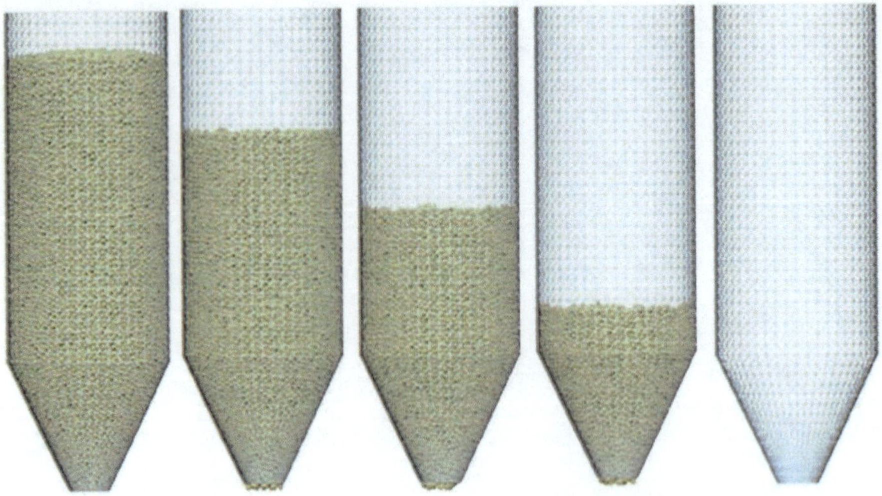

Fig. 6.4 Silo discharging process

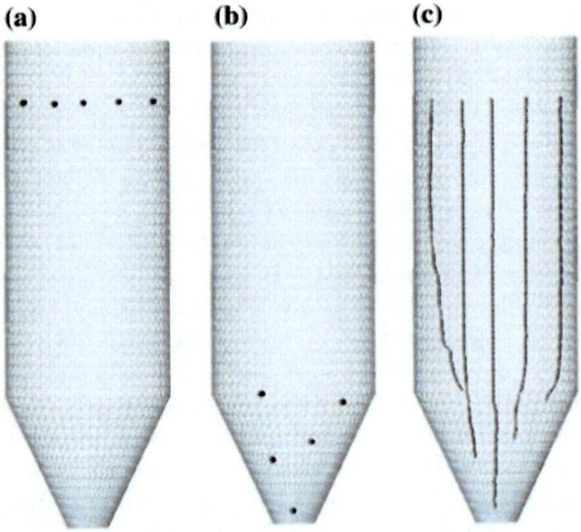

Fig. 6.5 Trajectory of five marked points

With the flow of particles, the closer to the unloading port, the more obvious the extrusion and friction. The resistance along the flow direction also increases and reaches the maximum at the discharge port, but at this time, the resistance of the coal particles in the upper part of the silo is smaller and the flow speed is faster, resulting in additional pressure to compact the coal under the silo, and the coal near the silo wall is subjected to greater friction resistance which makes the flow more difficult.

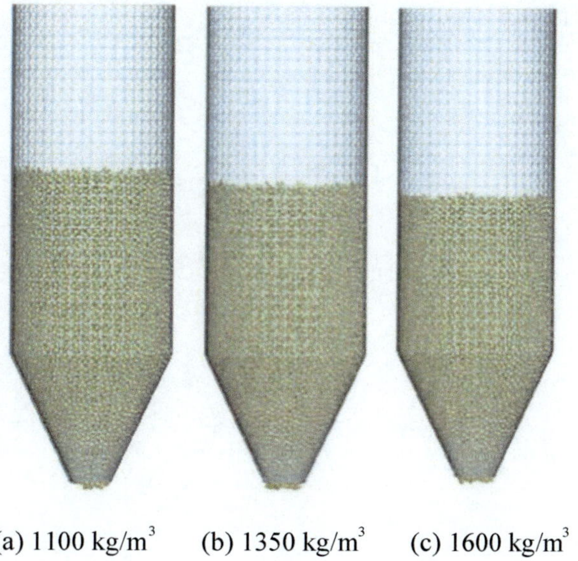

Fig. 6.6 Flow pattern diagram of different densities silo discharge

6.2.2 *Factors Affecting Flow Status*

6.2.2.1 The Density of Particles

In the silo, in order to study the influence of the density of the coal particles on the flow state, when the height of the silo was 18 m, the diameter of the silo was 6 m, the inclination angle of the discharge hopper was 65°, the opening radius of the tapered discharge hopper was 0.75 m, the particle size was 0.1 m, the coefficient of static friction between the coal particles was 0.45, and coefficient of static friction between the coal particles and the wall was 0.35, the density of coal was selected as 1100, 1350 and 1600 kg/m^3, respectively. By observing the flow state under different conditions, it is convenient to find out how the particle density affects the flow state of coal particles. The flow condition at $t = 70$ s is shown in Fig. 6.6.

From the whole unloading process and Fig. 6.6, it can be seen that the change of density of coal bulk has no obvious effect on the flow state of dry bulk in the coal silo, which is mainly manifested as the overall flow. However, during the unloading process, there are occasionally uneven hopper flow. Observing the whole unloading process, the unloading time is 133 s when the density of coal bulk is 1600 kg/m^3, the unloading time is 138.2 s when the density is 1350 kg/m^3 and the unloading time is 145.3 s when the density is 1100 kg/m^3, so the higher the particle density in the silo, the faster is the unloading speed.

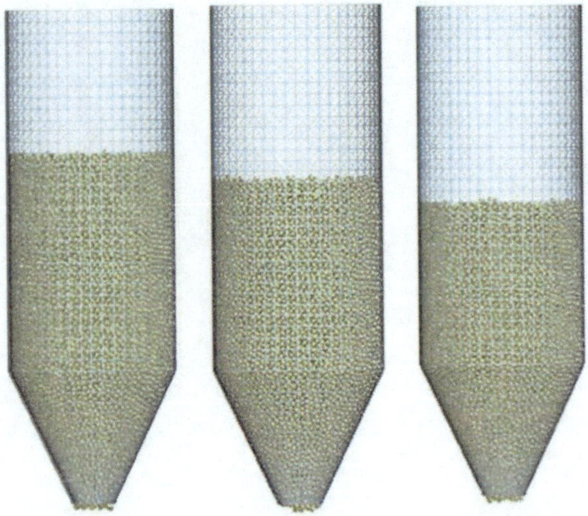

Fig. 6.7 Flow pattern diagram of different particle sizes distribution silo discharge

6.2.2.2 Distribution of Particle Size

When studying the influence of particle size distribution on the flow state of coal particles in the silo, the height of the silo was 18 m, the diameter was 6 m, the inclination angle of the tapered unloading hopper was 65°, the opening radius of the discharge hopper was 0.75 m, the density of coal was 1600 kg/m^3, the coefficient of static friction between the coal particles was 0.45, the coefficient of static friction between the coal particles and the wall was 0.35, and the particle size distribution was 0.08, 0.09 and 0.1 m, respectively. Figure 6.7 shows the flow pattern at $t = 70$ s. The particle sizes of coal from left to right are 0.08, 0.09 and 0.1 m, respectively.

It can be seen from the whole unloading process and Fig. 6.7 that the flow state of the selected particles with larger volume is basically the same during unloading, which basically belongs to the whole flow. The unloading time of particles with 100, 90 and 80 mm is 133, 134.3 and 136 s, respectively. To sum up, when other conditions are the same, the larger the particle size distribution of coal bulk, the faster is the unloading speed.

6.2.2.3 Coefficient of Static Friction Between Coal Particles

When studying the influence of the coefficient of static friction between coal particles on the flow state in the silo ensured that the height of the silo was 18 m, the diameter was 6 m, the inclination angle of the tapered unloading hopper was 65°, the opening radius of the discharge hopper was 0.75 m, the density of coal was 1600 kg/m^3, the particle size distribution was 0.1 m and the coefficient of static friction between the coal particles and the wall was 0.35. The coefficients of static friction between the coal particles were 0.45, 0.3 and 0.15, respectively. Figure 6.8 shows the discharge flow pattern at $t = 70$ s.

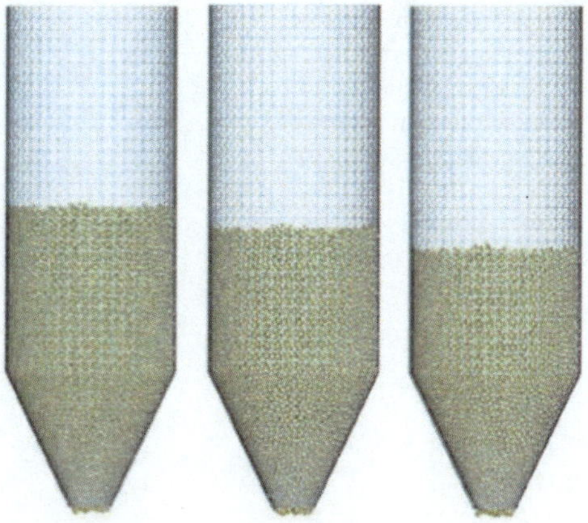

Fig. 6.8 Flow pattern diagram of different coefficients of static friction between the coal particles silo discharge

From the whole unloading process and Fig. 6.8, it can be seen that under other conditions being the same, the smaller the coefficient of static friction between the coal particles, the more stable and smooth the flow state of particles, the closer to the overall flow. When the coefficient of static friction between the coal particles is 0.45, the unloading time is 133 s; when the coefficient of static friction is 0.3, the unloading time is 124 s; when the coefficient of static friction is 0.15, the unloading time is 112.8 s, indicating that the smaller the internal friction coefficient, the faster is the unloading speed. The kinetic energy of particles is lost due to friction during unloading, so the larger the coefficient of static friction, the greater will be the kinetic energy loss, and the slower will be the unloading. Therefore, reduce the coefficient of static friction between the bulk materials, so that the unloading speed of particles will be faster.

6.2.2.4 Coefficient of Static Friction Between the Coal Particles and the Wall

When studying the influence of the coefficient of static friction between the coal particles and the wall on the flow state of the coal bulk, when the height of the coal silo was 18 m, the diameter was 6 m, the inclination angle of the tapered unloading hopper was 65°, the opening radius of the discharge hopper was 0.75 m, the bulk density of coal was 1600 kg/m^3, the particle size distribution was 0.1 m, and the coefficients of static friction between the coal particles is 0.45; the coefficient of static friction between the coal particles and the wall were 0.35, 0.25 and 0.15 at $t =$ 70 s, respectively.

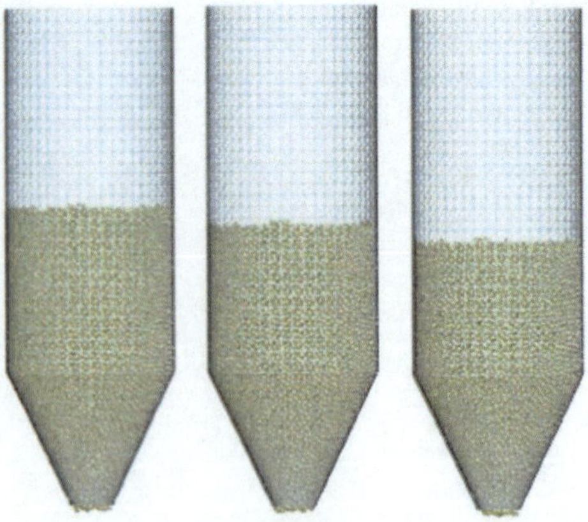

Fig. 6.9 Flow pattern diagram of different coefficients of static friction between the coal particles and the wall silo discharge

From the whole unloading process and Fig. 6.9, it can be seen that when other conditions are the same, the smaller the coefficient of static friction between the coal particles and the wall, the closer the flow state of the coal bulk is to the overall flow, the greater the coefficient of static friction is, and the closer it is to the central flow, especially the flow velocity of the coal bulk near the silo wall is smaller. When the coefficient of static friction is 0.25, 0.35 and 0.15, the unloading time is 133, 125 and 116.2 s, respectively, so the smaller the coefficient of static friction is, the faster the unloading speed is.

6.2.2.5 Inclination Angle of Conical Discharge Hopper

In the study of the influence of the inclination angle of the unloading hopper on the flow state of the coal bulk in the coal silo, the height of the coal silo was 18 m, the diameter of the coal silo was 6 m, the opening radius of the conical unloading hopper was 0.75 m, the density of the coal bulk was 1600 kg/m^3, the coefficient of static friction between the coal particles was 0.45, the coefficient of static friction between the coal particles and the wall was 0.35, and the inclination angles of the conical unloading hopper were 45°, 55° and 65°, respectively.

It can be seen from the whole unloading process and Fig. 6.10 that, under the same other conditions, the larger the inclination of the conical unloading hopper, the closer is the bulk flow state to the overall flow during unloading. When the taper angle is 65°, 55° and 45°, the unloading time is 133, 164 and 187.6 s, respectively. Therefore, the larger the inclination of the conical discharge hopper, the faster the discharge speed of the coal bulk.

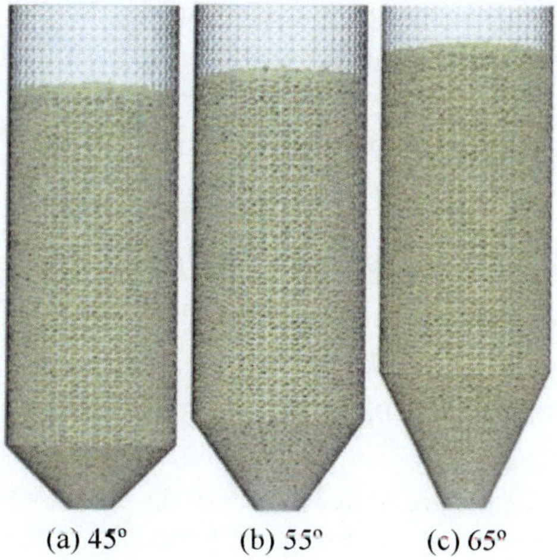

Fig. 6.10 Flow pattern diagram of different inclination silo discharge

6.2.2.6 Opening Size of Tapered Discharge Hopper

When studying the influence of the opening size of the tapered discharge hopper on the flow state of the coal bulk in the coal silo ensured that the height of the coal silo was 18 m, the diameter of the coal silo was 6 m, the inclination angle of the conical discharge hopper was 65°, the density of the coal bulk was 1600 kg/m^3, the coefficient of static friction between the coal particles was 0.45, the coefficient of static friction between the coal particles and the wall was 0.35, and the opening radius of the tapered discharge hopper was 0.7, 0.75 and 0.8 m, respectively.

From the whole unloading process and Fig. 6.11, it can be seen that when other conditions are the same, the larger the opening size of unloading, the closer the

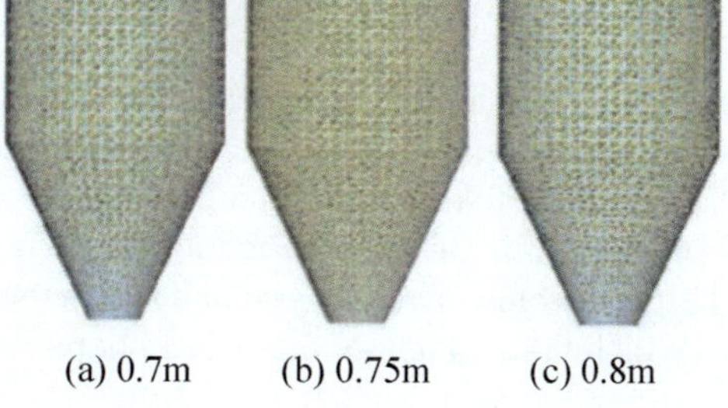

Fig. 6.11 Flow pattern diagram of different opening sizes silo discharge

bulk flow state is to the overall flow. When the opening radius is 0.7, 0.75 and 0.8 m, respectively, the corresponding unloading time is 163.3, 133 and 112 s. Therefore, the larger the opening size of the conical discharge hopper, the faster is the discharging speed of the coal bulk.

6.3 Simulation of DEM for Bulk Material Flow in Coal Silo Unloading Hopper

6.3.1 Hopper Discharge Theory

The hopper is the key part of the coal storage silo during unloading, and most of the traditional hoppers in the lower part of the silo are designed to be linear conical or pyramidal (square conical), but according to the actual production and use of coal mines in the past, these two types of hoppers often have accidents of arching and blocking the silo; especially when the bulk particles of coal storage are viscous, the water content is high and there are too many very fine particles, the above problems are especially obvious. Therefore, it is urgent to improve the design of hopper structure.

For the blockage problem in the unloading process of the traditional conical hopper, the main reasons are attributed to the hopper structure and the characteristics of coal particles. In view of the various degrees of damage of coal particles in the process of loading, unloading and transportation, a large number of fine particles are produced. Because of the different fluidity of coarse and fine particles and the penetration of fine particles, when the particles flow slowly in the hopper, it is easy to cause the phenomenon of bulk segregation. The fine particles are concentrated in the center of the hopper and flow out first, and the coarse particles are concentrated around the hopper and finally discharged. This can easily lead to arch effect, suspended material and even blockage and so on. In order to eliminate the blockage, workers often take manual beating, steel brazing materials and other arch-breaking methods, but in the long run, they will damage the hopper strength, and even cause deformation and so on.

In view of this, referring to the relevant application experience of mining, hydropower and metallurgical departments, the conical silo hopper is improved into a circular hyperbolic hopper, its cross-section is shown in Fig. 6.12, the solid line is the hyperbolic hopper, the dashed line is the conical hopper and the cross-section is circular.

Compared with the conical hopper, the hyperbolic hopper has two remarkable characteristics in structure: the first is the equal cross-section shrinkage, and the cross-sectional shrinkage of the conical hopper increases gradually from top to bottom. The second is that the inclination angle of the hopper wall is different at different heights, which increases gradually from the loading port to the unloading port, while the inclination angle of the conical hopper wall is a constant value.

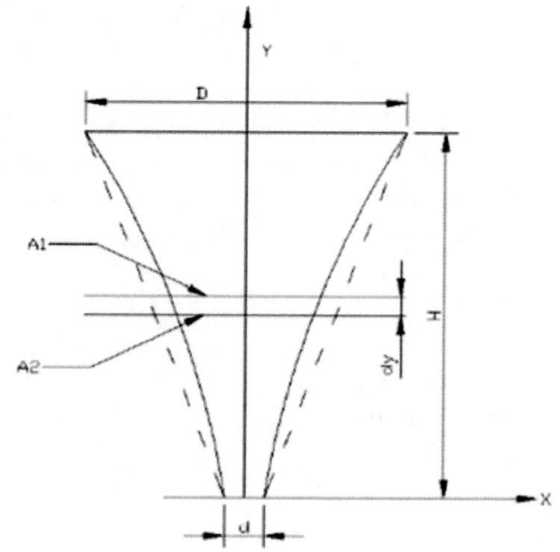

Fig. 6.12 Hyperbolic hopper

During unloading, the particles in the hopper move slowly downward to overcome the friction resistance between the particles and between the particles and the hopper wall under the action of their own gravity. When the particles in the hopper flow from section A_1 to section A_2, the decrease of cross-sectional area makes the extrusion between particles more serious and the position between particles is redistributed. The reduction of the cross-section is expressed by the section shrinkage *C*. According to the actual test, the larger the section shrinkage *C*, the more severe is the particle extrusion of the next section, and the greater is the resistance, and the more disadvantageous it is to unload. When the resistance is equal to the gravity of the particle itself, the particle stops flowing downward, and then the warehouse plugging accident occurs. It is generally believed that increasing the inclination angle of the hopper cone to make it larger than the friction angle between the coal bulk and the inner wall of the silo can reduce the wall friction of particles and contribute to the flow of particles. The decrease of cross-section is obviously not conducive to unloading, but the inclination angle of hyperbolic hopper wall is different at any height closer to the discharge port, and the larger is the inclination angle, which to a certain extent makes up for the resistance increment between particles caused by the decrease of cross-section. As a result, the resistance of particles from top to bottom is basically constant, particles show an approximately uniform overall flow in the hopper, and there will be no accidents such as 'arch', 'suspended material' or even blockage.

6.3.1.1 Determine the Curve Equation of the Hyperbolic Hopper

This formula is calculated according to the shrinkage rate of the equal section.

$$\frac{A_1 - A_2}{A_1 dy} = \frac{dA}{A_1 dy} = KC \tag{6.1}$$

where dA is the reduction of cross-sectional area; A_1 and A_2 are the cross-sectional area; C is the equal section shrinkage; K is the shape factor of hopper section (circular: $K = 1$; square: $K = 0.75$–1); dy is a small amount of decline.

In the hopper with circular cross-section in this section,

$$\frac{dA}{A_1 dy} = \frac{2\pi x dx}{\pi x^2 dy} = KC \Rightarrow \frac{dx}{x} = \left(\frac{KC}{2}\right)dy \tag{6.2}$$

After calculus:

$$\int_{\frac{d}{2}}^{x} \frac{dx}{x} = \int_0^y \frac{KC}{2} dy \tag{6.3}$$

After simplification:

$$x = \frac{d}{2} e^{\frac{KC}{2} y} \ or \ y = \frac{2}{KC} \ln \frac{2x}{d} \tag{6.4}$$

6.3.1.2 Calculate the Shrinkage of the Cross-Section

Perform calculus on the formula (6.2) again.

$$\int_{\frac{d}{2}}^{\frac{D}{2}} \frac{dx}{x} = \int_0^H \frac{KC}{2} dy \tag{6.5}$$

Get:

$$C = \frac{2}{KH} \ln\left(\frac{D}{d}\right) \tag{6.6}$$

6.3.1.3 Calculate the Inclination of the Hopper Wall

Obtain from the transformation of the formula (6.2):

$$\frac{dy}{dx} = \frac{2}{KCx} = \tan\alpha$$

$$\tan\alpha = \frac{2}{KCx} \Rightarrow \alpha = \arctan\frac{2}{KCx} \tag{6.7}$$

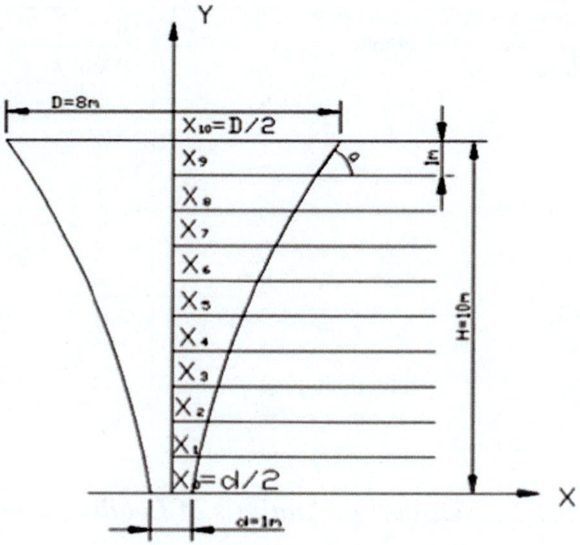

Fig. 6.13 Schematic diagram of the section of hyperbolic hopper

According to the hyperbolic hopper model of a coal preparation plant, the inner diameter of the upper and lower openings of the hopper are 8 and 1 m, respectively, the hopper height is 10 m ($H = 1.25d$), and the section is circular ($K = 1$).

According to the formula (6.6), the relevant values are substituted, and the cross-section shrinkage C is 0.415. In order to find out the inclination angle at each height of the hopper and the corresponding radius of curvature, the hopper height is divided into ten sections averagely, and each section is 1 m high, as shown in Fig. 6.13 in detail.

On each section, the corresponding inclination angle α and curvature radius x are calculated according to the formulas (6.7) and (6.4). The specific calculation results are shown in the Table 6.5. It can be seen from the table that the inclination angle of the hyperbolic hopper wall reaches the maximum at the discharge outlet, and the inclination angle of the conical hopper is constant $70°42'36''$ under the same conditions. Although the inclination angle of the upper part of the hyperbolic hopper is small, the extrusion of the particles in the lower part is more serious near the outlet, and the increased inclination angle can partially counteract the resistance between the particles, which is convenient for unloading.

Table 6.5 The segment parameters of hyperbolic hopper

y/m	x/m	$\alpha/°$	C
10	4.000	50°25′51″	0.415
9	3.236	56°07′04″	0.415
8	2.630	61°22′46″	0.415
7	2.137	66°05′10″	0.415
6	1.736	70°11′02″	0.415
5	1.411	73°40′45″	0.415
4	1.150	76°36′56″	0.415
3	0.932	79°03′23″	0.415
2	0.757	81°04′13″	0.415
1	0.615	82°43′25″	0.415
0	0.500	84°04′35″	0.415

6.3.2 *Simulation Analysis of Loading and Unloading*

6.3.2.1 The Establishment of the Model

Referring to the hopper model in the above example, three kinds of hopper models were designed: hyperbolic hopper, conical hopper and square hopper, with an equal wall thickness of 0.2 m. The cross-sectional area of the open outlet of the three kinds of hoppers is equal, in which the cross-section of the square hopper is square and the inner diameter is equal to its side length. The specific geometric structure parameters are shown in Table 6.6.

In order to compare the unloading process of hoppers with different structures, the following related parameters were set equally. Use the Pro/E to establish their respective models which was imported into Hypermesh for mesh division and numbering processing. The mesh types were all triangular elements, of which the size was 250 mm and the number was: hyperbolic hopper (1–5002); conical hopper (1–5114); square hopper (0–6096). After adding the material parameters of the warehouse wall,

Table 6.6 Geometry parameters of the three kinds of hopper

Parameters	Hyperbolic hopper	Conical hopper	Square hopper
Opening inner diameter/m	8	8	7.1
Outlet inner diameter/m	1	1	0.89
Height/m	10	10	10
The changing trend of inclination angle	Gradually increase	Constant	Constant
The changing trend of section shrinkage	Constant	Gradually increase	Gradually increase

it was saved as a.msh format file, which was imported into EDEM for discrete element unloading simulation.

The particle model in the coal silo uniformly choose round particles (average radius was 80 mm). The particle size is normally distributed with a variance of 0.05. Other particle parameters are shown in Tables 6.1 and 6.2.

6.3.2.2 Boundary Conditions, Loading and Unloading Simulation

There are two kinds of contacts in the model: contact between the coal particles, and contact between the coal particles and the silo wall. During the actual unloading, the coal particles are broken and deformed, and the silo wall also has micro-deformation. The two kinds of contact should be soft contact, that is, the two contact bodies are deformed in order to better reflect the actual situation. The contact between the actual particles and between the particles and the silo wall is complex. Here, in order to simplify the calculation, the two kinds of contact were simplified to adopt the Hertz-Mindlin (no slip) contact model. Took the acceleration of gravity as 9.81 m/s^2. The generation mode of particle factory was dynamic mode. The minimum computing grid radius was 2.6*R*, that is, 87412 grids. The target iterative data saving interval was 0.05 s and the time step was 0.002 s.

After completing the above relevant settings, a circular baffle was added to the discharge port of the three hoppers, and the discharge port is blocked when the particles are generated. The modeling process of the baffle was the same as that of the hopper, and then imported into EDEM. The particle generation stage was from 0 to 6 s, and a circular particle factory with a diameter of 3.4 m was set at 0.3 m below the hopper loading port. The formation rate of particles here was 5000 per second, a total of 30000. After standing for 4 s, the particles in the coal silo basically reached a stable state. Starting from 10 s, the circular baffle of the material outlet was removed, and the particles moved downwards under the action of their own gravity, overcoming the frictional resistance. Figure 6.14 shows the respective loading and unloading process of the three hoppers.

6.3.2.3 Result Analysis

When loading, the particles with initial velocity accelerate to the bottom of the hopper under the action of gravitational acceleration. As the cross-section of the hopper shrinks, the cross-sectional area decreases gradually, and the particles on both sides first contact with the silo wall, and the velocity of the particles decreases to the initial velocity. At the same height, the wall of hyperbolic hopper is closest to the central axis of hopper, followed by conical hopper. In the square hopper, the friction contact between the particles at the junction of the adjacent coal silo wall is relatively serious, and the downward movement speed of the particles is also slow, which is not conducive to loading. It can be seen that the cone structure hopper is the most suitable for loading.

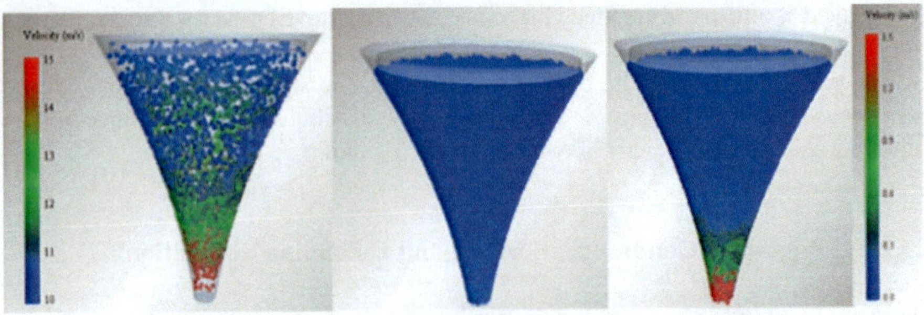

(a) Hyperbolic hopper

(b) Square hopper

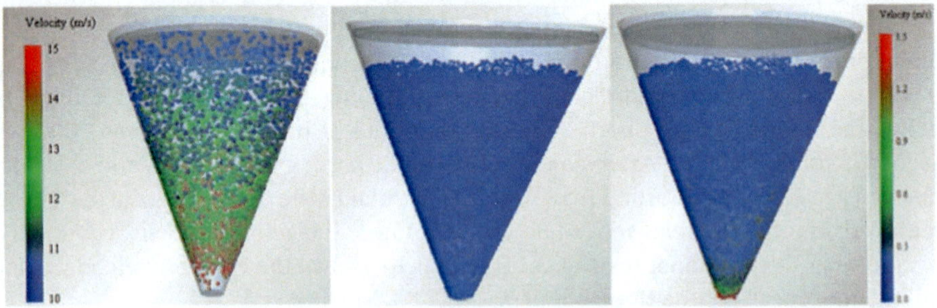

(c) Conical hopper

Fig. 6.14 The processes of loading and unloading of three kinds of hopper (loading, static, discharging)

When standing still, it is obvious that under the same conditions, the square hopper can be loaded the most, followed by the conical hopper, and the hyperbolic hopper can be loaded the least.

During unloading, it can be seen from the discharge figures of the three hoppers that the flow velocity at the discharge port of the hyperbolic hopper is obviously higher than that of the other two hoppers, followed by the square hopper and the conical hopper is the slowest. Under the same conditions, the numerical calculation of hopper discharge shows that all the particles in the conical hopper flow out at

109.25 s, the particles in the square hopper run out at 103.5 s, and when the particles completely flow out from the hyperbolic hopper, $t = 76.75$ s only taking 66.75 s, which is 32.5 s less than that of the conical hopper.

In addition, during unloading, the particles in the three kinds of hopper all show the flow form of hopper flow, the hyperbolic hopper is relatively more prominent, the central particle velocity is fast, and the particles flow out first, and only the particles in a certain range above the discharge port show this form. At the initial stage after the baffle is removed, the particles near the upper part of the loading port still flow as a whole.

References

1. Cannavacciuolo A, Barletta D, Donsi G, Ferrari G, Poletto M (2009) Arch-Free flow in aerated silo discharge of cohesive powders. Powder Technol 191(3):272–279
2. Liu, Y. (2013). Analysis on application effects of air cannon in blocked coal bunker. *TECHNO SCI COAL.*, 41(S2), 398–399. (in Chinese)
3. Zhu J, Peng F, Guo P, Zhu C, Wen X (2012) Study on testing for transient impact of air cannon for unclogging coal bunker. Mining & Processing Equipment. 40(06):107–109 (in Chinese)
4. Colonnello C, Reyes LI, Clement E, Gutierrez G (2014) Behavior of grains in contact with the wall of a silo during the initial instants of a discharge-driven collapse. Phys A 398:35–42
5. Mellmann J, Hoffmann T, Fuerll C (2014) Mass flow during unloading of agricultural bulk materials from silos depending on particle form, flow properties and geometry of the discharge opening. Powder Technol 253(46):52
6. Lin P, Zhang S, Qi J, Xing YM, Yang L (2015) Numerical study of free-fall arches in hopper flows. Phys A 417:29–40
7. Parafiniuk P, Molenda M, Horabik J (2013) Discharge of rapeseeds from a model silo: Physical testing and discrete element method simulations. Comput Electron Agric 97:40–46

Chapter 7
Stress Analysis of Silos Using DEM

For the side pressure calculation of the silo wall, when unloading is the core problem of the design of the silo structure [1], the design specifications of the silo structure in many countries are based on the further revision of the Janssen formula [2]. Due to the lack of further consideration of the influence of the storage flow in the silo, it cannot be used to accurately calculate the discharge pressure of the storage material.

With the further development of computer technology and simulation technology, many numerical methods for analyzing the pressure on the side of the silo wall have emerged [3, 4], and most researches have focused on the more common analysis of DEM [5, 6] and the numerical calculation of FEM [7–9]. The storage material is made up of discrete particle system, and the silo wall is constructed from a continuous geometric solid model. The two methods are different in emphasis. DEM can be traced to the mechanical condition of a single particle, and by applying a reasonable gravity field, it can make a better simulation of the whole process of loading and unloading material [10], but for such continuous media as the silo wall, DEM has great limitations, and it is difficult to simulate the calculation of its microscopic deformation. However, it is possible to calculate the stress and deformation of the silo wall more accurately using FEM. Therefore, when the problem of high nonlinearity such as the contact between storage material and silo wall is studied, these two numerical analysis methods can be combined for analysis.

In the silo discharge, the side pressure distribution of the silo wall is quite different from that of static, so it is necessary to treat the structure of the silo differently. The Janssen formula, the Ketchum theory, the Reimbret and Jenike calculation theory are more famous in many studies of silo pressure. Janssen formula and Ketchum theory are more suitable for static side pressure calculation for silo wall. On the basis of the existing static pressure theory, Reimbret and Jenike theory have made some corrections, which can be used as the theoretical basis for calculating the side pressure of the silo wall when the silo discharges. However, in the actual discharge

X. Wang et al., *Engineering Applications of Discrete Element Method*, Engineering Applications of Computational Methods 4,
https://doi.org/10.1007/978-981-15-7977-6_7

test, the difference between the test results and the theoretical prediction value is large, of which the specific reasons are still being explored.

When unloading, the problem of silo overpressure has always been the key to the study [11], and it is generally believed that the fundamental reason is the increase of the dynamic load effect coefficient caused by sudden disturbance of the storage balance [12, 13]. Available literature has shown that discharge increases side pressure volatility, exacerbates silo instability and makes it more prone to accidents [14]. In the actual discharge process, there should be a relationship between deformation expansion and bunker wall overpressure that occurs on the failure silo.

7.1 Silo Boundary Conditions and Loading and Unloading Material Simulation

Because EDEM has high requirements for computer hardware and equipment, in order to shorten the simulation time, we need to pay attention to the rationality of parameter design. In order to facilitate discharge, the radius of the discharge port of the silo should not be less than ten times the radius of the coal particles. In this paper, there were three particle sizes, with an average radius of 0.02, 0.05 and 0.08 m, respectively, and 40000, 40000 and 50000 particles were generated in a normal distribution form, with the set variance being 0.05, and the detailed parameters of coal are shown in Tables 7.1 and 7.2.

Table 7.1 Material parameters for coal particles

Particulate materials	Poisson's ratio	Shear modulus	Density
Coal	0.25	3×107 Pa	1300 kg/m^3

Table 7.2 Contact Parameters

Contact parameters	Value
Coefficient of restitution between the coal particles	0.1
Coefficient of static friction between the coal particles	0.45
Coefficient of rolling friction between the coal particles	0.1
Coefficient of restitution between the coal particles and the wall	0.15
Coefficient of static friction between coal particles and the wall	0.5
Coefficient of rolling friction between the coal particles and the wall	0.15

The particle factory generated three particles at a rate of 5000 per second. After the mixture was even, it was loaded into the silo. In the direction of the center of the silo, the gravity acceleration was set at 9.81 m/s^2, the operation time step was set to 0.002 s in the EDEM solver module, the minimum computing grid radius was set to $3R$ and the calculation data save interval was 0.05 s. The total simulation time was 142 s, and the time used to generate and mix particles and pre-preparation was 0–46 s.

Before starting the loading, a circular baffle with the diameter of 0.8 m was installed at the discharge port of the silo. The baffle needed to be left for two seconds after the loading was completed. Let it stand for a period of time to restore the stability of the coal bulk in the silo. At 48 s, the baffle was pulled away. At this time, the coal bulk in the silo gradually fell and flowed out of the silo under the influence of gravity. Figure 7.1 shows the velocity cloud of the particles in the coal silo when the coal silo is standing and discharging. It can be seen that the closer to the discharge port, the faster the coal bulk.

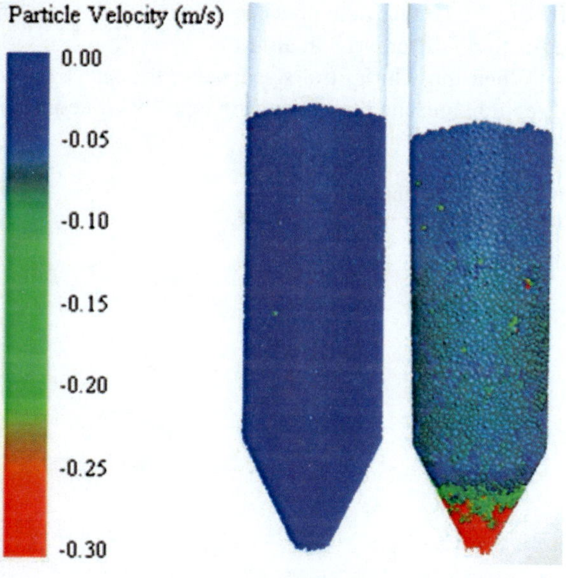

Fig. 7.1 Velocity cloud diagram of coal bulk

7.2 Simulation Analysis of Static Dynamic Pressure

7.2.1 The Classic Theory of Coal Barn Pressure

7.2.1.1 Straight Silo Pressure Distribution

In ordinary liquid container, according to the Pascal principle and liquid pressure calculation formula, it can be seen that the pressure produced by the liquid is proportional to the depth, while the pressure at the same horizontal position is equal lying large. Although coal bulk shows a certain fluid nature in the flow process, the pressure of coal bulk in coal silos is not exactly the same as the liquid pressure. For this situation, Janssen used the differential balance equation to establish a mathematical model of the silo wall micro-unit, and by calculation, derived the silo wall static stress formula of the material in the straight silo, that is, the Janssen theory formula. The theory needs to be established with the following assumptions: (1) Incompressible, particulate materials of the same nature are evenly filled in the silo; (2) the vertical pressure of the particles in the same horizontal surface is equal in size; (3) the ratio of vertical pressure in the silo to the horizontal pressure remains unchanged at any point. Accordingly, the coefficient of friction inside the particle is a constant. Figure 7.2 is a particle pressure schematic in the straight silo, taking one of the thin units to do differential analysis.

When the silo is in a static state, the sheet is balanced by force in the vertical direction, and the corresponding equilibrium equation is:

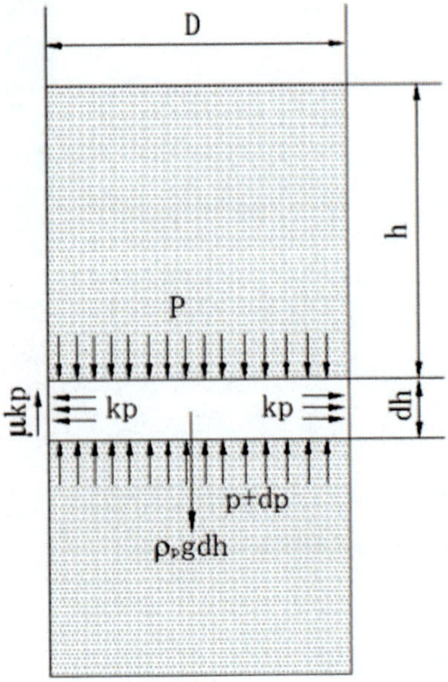

Fig. 7.2 Particle pressure schematic in a straight silo

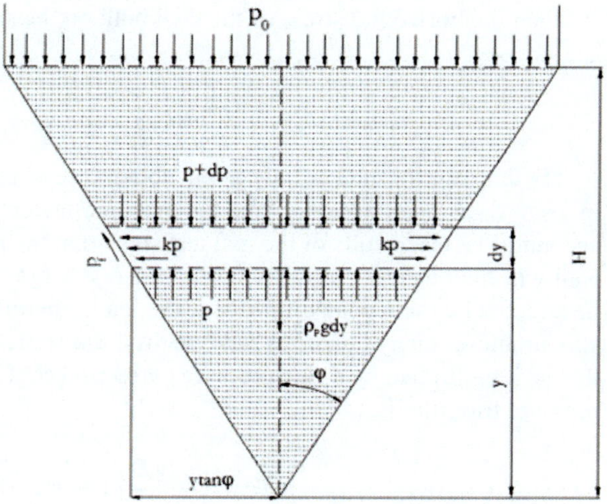

Fig. 7.3 Hopper chamber particle pressure schematic

$$\frac{\pi}{4}D^2p + \frac{\pi}{4}D^2\rho_p gdh = \frac{\pi}{4}D^2(p + dp) + \pi D\mu kpdh \tag{7.1}$$

where D is the silo diameter; p is the pressure at this depth, reference Fig. 7.3; g is the gravity acceleration; ρ_p is the particle density in the silo; μ is the coefficient of internal friction between the particles and the wall of the silo; k is the side pressure coefficient, which is the ratio between the horizontal stress p_h of the particles in the silo and the vertical stress p_v.

When the internal friction angle φ is fixed, according to the Coulomb formula, the calculated side pressure coefficient of the scatter destruction envelope can be obtained as follows:

$$k = \frac{1 - \sin\phi}{1 + \sin\phi} \tag{7.2}$$

Substituting formula (7.1), and sorting can be obtained:

$$\left(D\rho_p g - 4\mu kp\right)dh = Ddp \tag{7.3}$$

Integrate the left and right sides separately to get:

$$h = -\frac{D}{4\mu k}\ln\left(\rho_p g - \frac{4\mu kp}{D}\right) + \frac{D}{4\mu k}\ln\left(\rho_p g\right) \tag{7.4}$$

The particle vertical stress of the Janssen formula is:

$$p_v = p = \frac{\rho_p gD}{4\mu k}\left[1 - e^{\left(-\frac{4\mu kh}{D}\right)}\right] \tag{7.5}$$

Then the horizontal stress of the coal bulk on the silo wall is $p_h = kp_v$, that is

$$p_h = \frac{\rho_p g D}{4\mu}\left[1 - e^{(-\frac{4\mu k h}{D})}\right] \tag{7.6}$$

The Janssen formula is only applicable to pressure calculations when the silo is in a static state and does not take the dynamic factors in the discharge process into account. The test results of the real test show that the pressure on the side of the silo wall will fluctuate significantly during the discharge process, and the side wall will be affected by the repeated impact of the load. Therefore, for the design standard of silo structure, many countries have revised the static pressure calculation formula, that is, it multiplied by a side pressure correction coefficient, and the corrected static pressure formula is:

$$p_v = C_v \frac{\rho_p g D}{4\mu k}\left[1 - e^{(-\frac{4\mu k h}{D})}\right] \tag{7.7}$$

$$p_h = C_h \frac{\rho_p g D}{4\mu}\left[1 - e^{(-\frac{4\mu k h}{D})}\right] \tag{7.8}$$

where C_v and C_h are the side pressure correction coefficients in the vertical direction and the horizontal direction, respectively.

If $h \to \infty$, the vertical pressure at that time is

$$p_v = \frac{\rho_p g D}{4\mu k} \tag{7.9}$$

Therefore, for a certain volume of coal silo, when the coal bulk in the silo filled to a certain extent, the static pressure in the silo will approach a fixed constant.

7.2.1.2 Discharge Hopper Pressure Distribution

Similar to a straight silo, the bulk pressure in the discharge hopper of the coal silo can also be derived according to the differential unit method of the Janssen formula. Figure 7.3 is a diagram of the bulk pressure in the discharge funnel silo, taking a micro-unit body inside it, and performing a balance analysis on the vertical force.

Since the area of the hopper cross-section is gradually reduced along the vertical direction, the angle of the discharge hopper is 2ϕ, the pressure of the micro-unit layer in the figure along the vertical direction is p, the pressure along the horizontal direction is kp, the friction along the direction of the hopper wall is p_f, and the friction along the direction of the hopper wall is:

$$p_f = \mu p(k\cos^2\phi + \sin^2\phi)\frac{dy}{\cos\phi} \tag{7.10}$$

When the hopper is in a static state, the particles of the micro-unit layer are in equilibrium in the vertical direction, and the equilibrium equation is:

$$\pi(y\tan\varphi)^2\left[(p+dp)+\rho_p g dy\right]= \\ \pi(y\tan\varphi)^2 p+2\pi(y\tan\varphi)p_f\cos\varphi \tag{7.11}$$

Substituting formula (7.10), simplified available:

$$\frac{dp}{dy}=\frac{p}{y}\cdot\frac{2\mu}{\tan\phi}(k\cos^2\phi+\sin^2\phi)-\rho_p g \tag{7.12}$$

Analysis of the pressure distribution of coal silo discharge hopper under different circumstances:

(1) If thc coal silos arc only hopper silos, or if there is no coal in the straight silo above them, the boundary conditions of the coal silos should be: $y = H, p = 0$, the differential equation corresponds to the solution formula (7.12).

$$p=\frac{\rho_p g y}{\alpha-1}[1-(\frac{y}{H})^{\alpha-1}] \tag{7.13}$$

where $\alpha = 2\mu(kcos^2\phi + sin^2\phi)/tan\phi$, and $\alpha \neq 1$; if $\alpha = 1, p = \rho_p gyln(H/y)$.

(2) If the coal silo is a combination of hopper silo and straight silo, and there is a certain amount of coal bulk stored in the straight silo, then the corresponding boundary condition is $y = H, p = p_0$, and the solution formula (7.12) corresponds to the differential equation.

$$p=\frac{\rho_p g y}{\alpha-1}[1-(\frac{y}{H})^{\alpha-1}]+p_0(\frac{y}{H})^{\alpha-1} \tag{7.14}$$

where $\alpha \neq 1$; if $\alpha = 1, p = \rho_p gyln(H/y) + p_0 y/H$.

Through theoretical analysis, it is found that the pressure distribution of coal bulk in the discharge hopper is affected by the coal bulk in the upper straight silo. On the surface, the coal bulk in the silo only increases the pressure of the coal bulk in the hopper; in fact, the pressure distribution curve of the coal bulk will change fundamentally, but its overall trend will not change.

7.2.2 *Static and Dynamic Pressure Analysis*

After the simulation calculation was complete, the load parameters of coal bulk to the silo wall were exported from the EDEM data post-processing module. Figure 7.4 is a comparison of the trend of side pressure changes in the coal silos under static dynamic.

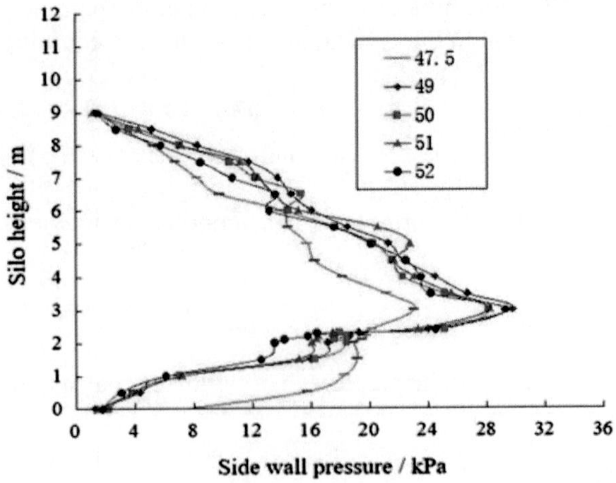

Fig. 7.4 The side pressure value of the silo wall for each point in static dynamic

When the discharge bezel is removed, the coal bulk at the bottom of the coal silo begins to fall under the force of gravity, destroying the equilibrium state maintained in the static silo, and the coal bulk in the middle level begin to slide, and then gradually pass edgy to the entire coal silo. The side pressure of the bunker wall produces a significant volatility change.

It can be seen that when the coal silo is standing, due to the shear and friction between the coal particles and between coal particles and the wall, with the increase of depth, the side pressure of the straight wall gradually increases, but the side pressure of the silo wall is decreasing, which is basically consistent with the theoretical side pressure value calculated according to the Janssen formula. The results of numerical calculation also found that the growth trend of static side pressure has certain fluctuating characteristics. In the process of discharge of coal silo, the motion state of coal bulk changes, resulting in the redistribution of the relative position between the particles. The coal bulk in the silo are along the silo wall at the same time downward sliding, presented as the overall flow of the form, resulting in the coal bulk to the silo wall friction squeeze enhanced. In the straight wall position side pressure increased, volatility increased, resulting in a greater peak side pressure. The side pressure value of the outlet position decreases rapidly. In general, regardless of whether the coal silo is at rest or unloaded, the side pressure peak of the silo wall occurs at the transition position of the cylinder and cone and is biased toward the side of the straight silo.

Through the calculation of side pressure numerical calculation, it has been found that during the discharge process, the flow of coal bulk in the silo increases the dynamic load coefficient, the side pressure has a certain degree of volatility at different times, the side pressure has a significant increase trend in the side wall of the straight silo and the hopper silo decreases rapidly. In this case, it is clear that the traditional numerical formula is no longer applicable, for safety reasons, the straight wall and hopper wall transition area of the silo wall can be appropriately reinforced.

7.3 Overpressure Discrete Element Simulation Analysis

From the previous analysis, it is known that the side pressure of the silo wall will increase significantly during discharge. Comparing the side pressure value corresponding to the initial four moments when the coal bin is standing and the initial discharge, the comparison of the overpressure coefficient is obtained, as shown in Fig. 7.5. It is found that in the moment of discharge, the position of the top of the silo is rapidly overpressure, and then the overpressure ratio decreases gradually with the outflow of coal bulk. In the straight silo site, with the increase in depth, the supercharge ratio has a certain easing, the overpressure ratio gradually decreased, and the increase of side pressure increased flatly. The dynamic balance between coal bulk is disturbed, causing the pressure on the side of the silo wall to fluctuate up and down, resulting in the overpressure coefficient of the straight wall side pressure that also fluctuates up and down, and the super-pressure peak appears in the upper middle of the straight silo. The overpressure ratio is as high as 1.73. In contrast, the side pressure of the hopper wall is greatly weakened, and as the depth increases, the overpressure coefficient decreases gradually, and the overpressure value is less than 1 at this time. In the moment of discharge, with the increase of silo depth, the overall overpressure law shows a certain linear change trend. With the increase of silo depth, the speed of coal bulk in the silo is gradually increasing, whether there is some kind of connection between the two is still to be further examined. In view of the problem of silo discharge overpressure, many scholars at home and abroad have done many experiments. It was found that from the top of the silo to a specific depth, with the change of material height, the change of boost coefficient has a certain law:

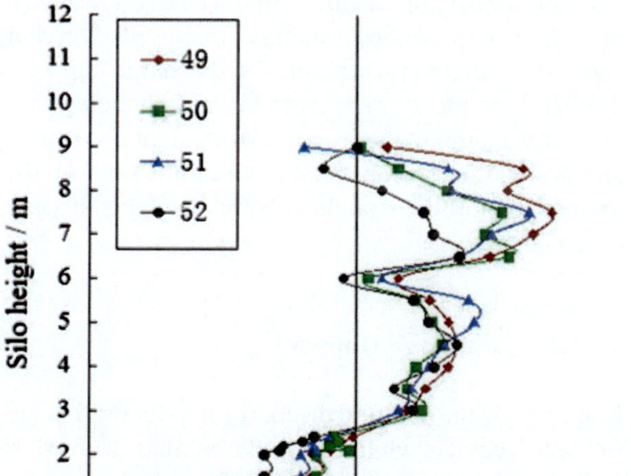

Fig. 7.5 Side pressure boost ratio for each point at four moments when unloading

the side pressure and the boost coefficient are gradually reducing with the outflow of particles in the silo.

In the discharge process, even if only a small amount of coal bulk flow, it will greatly affect the entire storage system. Then the storage in the silo as a whole began to move, the balance maintained when set aside is broken, resulting in a significant increase in pressure, which is the root cause of overpressure phenomenon. In order to avoid the occurrence of overpressure phenomenon, appropriate measures may be considered to prevent the overall flow of materials, and then from the top layer to start layer-by-layer discharge, the lower storage material remains stationary. In the moment of discharge, both the dynamic load coefficient and the instability factor of the coal silo increase, and the side pressure value of the silo discharge is usually expressed by the theoretical calculation of the static stress of the storage material multiplied by the overpressure coefficient. In order to eliminate other unstable factors, a suitable decompression tube can be installed at design time to reduce the increased side pressure when unloading, and by regulating the flow of storage material in the silo, thereby reducing the dynamic load coefficient at discharge.

7.4 Test Verification

7.4.1 Test Device

7.4.1.1 Overall Design

In order to verify the accuracy of the analysis results, the coal silo side pressure test is verified. The test device is mainly composed of discharge hopper, rectangular straight silo and auxiliary mechanism. The discharge hopper is responsible for regulating the discharge of the storage material and changing the size of the inclination of the discharge material, the rectangular straight silo is used to store and release coal bulk, and the auxiliary mechanism includes the bracket, the discharge plate, etc., which is used to support the straight silo and discharge hopper to ensure the stable operation of the device.

7.4.1.2 Discharge Hopper

Figure 7.6 is the pyramid discharge hopper used in the test and is the core component of the test device, which is composed mainly of two side panels. In the middle of the two-side plate 1 is set a transparent plate for the observation of coal bulk, another pair of side plate 2 is located in the middle of the two-side plate 1, and the fan plate is close to the side plate 1 to prevent leakage. The discharge plate is located just below the outlet, controlling the opening and closing of the outlet, and other supports and connecting plates play a supporting role.

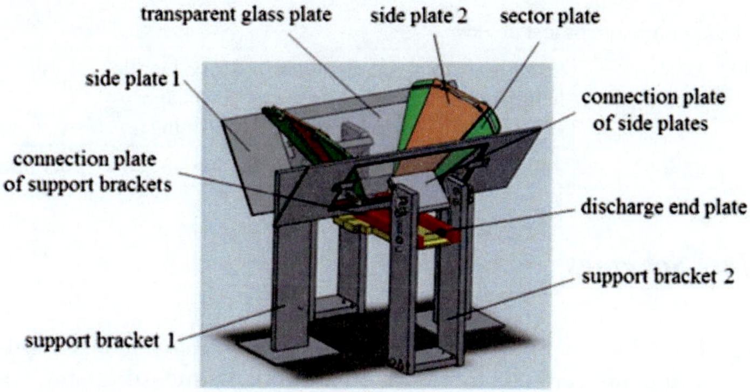

Fig. 7.6 Pyramid hopper

7.4.1.3 Rectangular Silo

Figure 7.7 is an assembly and physical drawing of the test device. The test device adopts a rectangular straight silo, which is composed of eight transparent glass plates, which is convenient to observe the movement of coal bulk. The rectangular straight silo is connected vertically with the discharge hopper, and the glass glue is coated at the connection to strengthen the bonding. The specific geometric parameters of the test device are shown in Table 7.3.

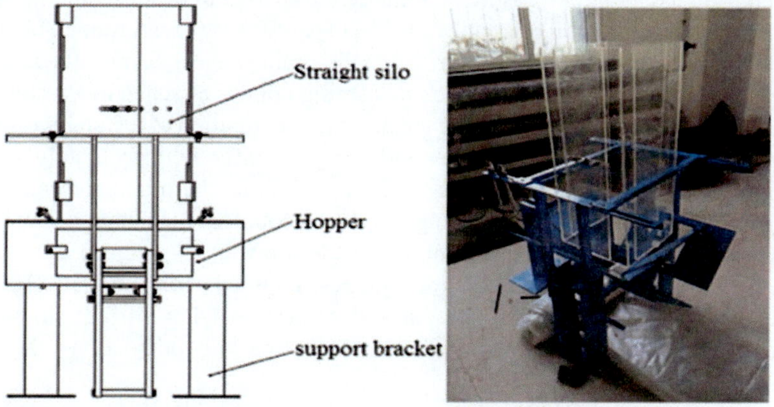

(a) Assembly drawing (b) Physical drawing

Fig. 7.7 Test device

Table 7.3 The parameters of test device

Silo height	Hopper inclination	Straight wall height	Side length of straight silo tope edge	Side length of discharge opening	Silo thickness
910 mm	45–65°	750 mm	200–300 mm	80 mm	3 mm

7.4.2 Test Schemes

In order to obtain the position of the silo wall side pressure distribution when the silo is placed and discharged, there are two common measurement methods for measuring the effective pressure of each point of the silo wall: direct measurement or indirect measurement. Direct measurement is the real-time monitoring of the right pressure sensor silos at each measuring point, and direct access to side pressure data on the silo wall. Although this method is simple, convenient and reliable, but the pressure sensor is usually thick, inconvenient to install on the test device, even if the pressure can be effectively measured, the resulting data will have a large error, not conducive to the results analysis. Indirect measurement is to use the principle and method of measuring strain by the appropriate strain gauge regularly on the points of the silo wall, to convert the deformation of the silo wall into the electrical signal of the strain gauge resistance change, and then use the strain tester to convert the received electrical signal into the silo wall pressure, so the method has a wide range of applications in the general model test field.

Therefore, the indirect measurement method was used in this test, and the sensor used in the test uses the foil resistance strain gauge of BX120-5AA model. In view of the symmetrical distribution of silo structure, the strain gauge is evenly attached to the silo wall as shown in Fig. 7.8, and each measuring point is attached to a strain gauge inside and outside the silo wall, where the pressure distribution of rectangular straight silo wall is tested. Limited by the test conditions, it is difficult to accurately measure the strain changes of each measuring point position, so the qualitative method of testing and analysis of each measuring point was adopted, focusing on monitoring the strain of the position slot in the middle of the silo wall.

Each point is fitted with two strain gauges, coupled with the DH3810 strain adjuster, connected with the tester via a semi-bridge dying wiring, suitable for measuring the bending strain of the silo body (the silo body along the radial displacement) (Fig. 7.9).

The tester uses the DH5923 dynamic signal test and analysis system containing the strain adapter and the supporting DHDAS_5923.V4.3.50 data acquisition and analysis system to record the strain change of the silo wall, the particle size of the coal used in the test is 6–13 mm, and the normal flow is guaranteed before the test. The specific operating procedures for the experiment are as follows:

(1) The strain gauge is distributed on the silo wall according to the position of Fig. 7.8, and the strain gauge is calibrated at the same time;

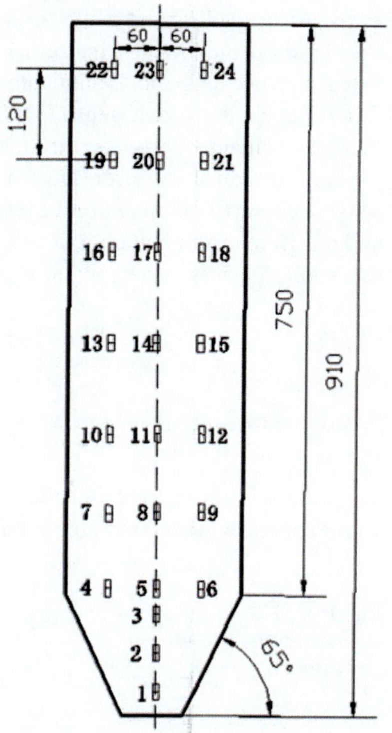

Fig. 7.8 Distribution of paste position of the strain gauge

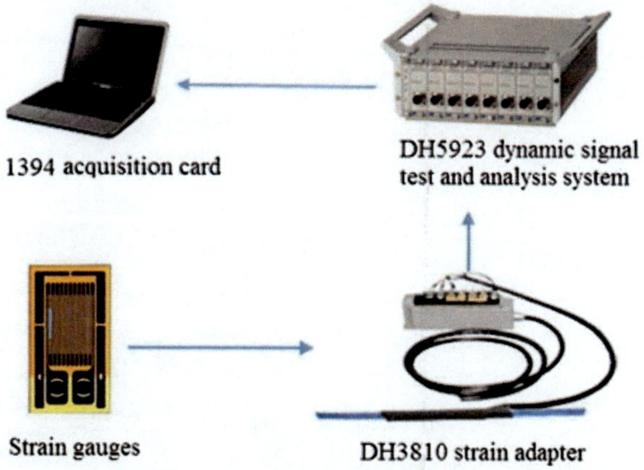

Fig. 7.9 The connection method for Instrument test

(2) Assembly test device, hopper inclination of 65°, according to the requirements of the matching test equipment and computer connection, and preliminary commissioning;

(3) After debugging the device, the coal bulk after drying will be loaded into the silo;

(4) After loading, set aside for a moment, wait for the coal to return to a stable state after opening the test instrument, and then pull out of the discharge bezel, begin to test, record the static dynamic under the silo wall strain;

(5) Repeat the above steps (3), (4), and conduct multiple tests.

In the actual test process, through multiple tests, the results of the test to average the data to reduce the error. Due to the limitation of the test conditions, it was not possible to carry out quantitative test, test the change of the pressure of the silo wall under different funnel inclination and different coal bulk particle size conditions, so the qualitative test measurement was carried out on some of the measuring points in the text.

7.4.3 *Analysis of Results*

The points in the middle of the silo in Fig. 7.8 were tested. After processing, the strain curves at each point are obtained as shown in Fig. 7.10.

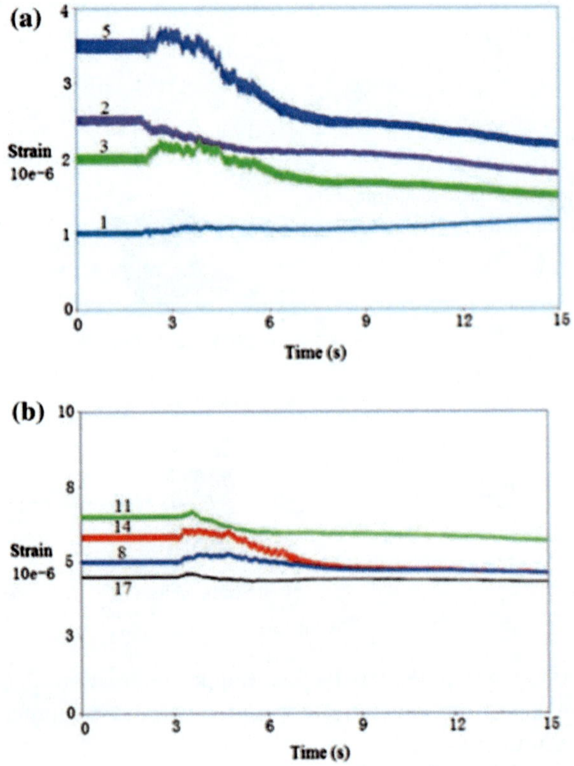

Fig. 7.10 The change of strain for every measure point from test device

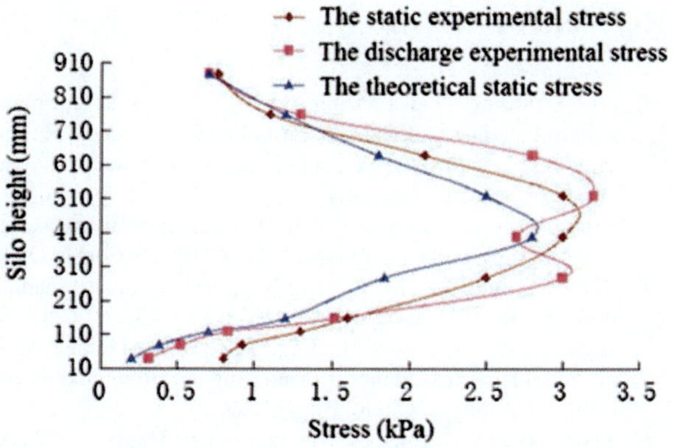

Fig. 7.11 Comparison graph of silo wall stress

It can be seen from Fig. 7.10 that the horizontal part on the left side of the curve corresponds to the silo static phase in the test. And when the discharge bezel is pulled away at about 3 s, the strain curve of the silo wall begins to fluctuate, the strain of the various measuring points on the straight silo (the measuring point 5–17) increases significantly, and then the silo wall strain decreases gradually with the discharge of the storage material, and has a certain degree of volatility, and tends to balance after all the coal bulk flows out. Because the flow rate of coal bulk at the hopper wall is large, the force of coal to the silo wall is reduced, so the strain of the measuring points (measure sitter 1–3) on the discharge hopper wall shows a certain downward trend in the discharge process. The experimental data measured at each point are collated and compared with the theoretical calculation values to obtain a graph of the wall stress comparison, as shown in Fig. 7.11.

It can be seen from Fig. 7.11 that the silo static pressure measured by test is basically the same as the distribution of the theoretical static pressure, while the silo wall of the straight silo has been different to varying degrees of overpressure. Influenced by the test error, there are some local differences in the pressure of the bunker wall side, but the overall trend is basically consistent with the analysis results of the study, and the side pressure of the funnel bunker wall is affected by the flow of coal bulk to a certain extent. The stress value measured in the test is the largest at point 11, corresponding to the silo height of 510 mm, and the pressure value is 3.2 kPa. According to the test results, in the silo static and discharge state, the maximum side pressure of the silo wall exists in the middle and lower part of the straight silo, and there is a significant overpressure phenomenon in the upper part of the silo when unloading, and the structure of the upper part of the silo can be considered for reinforcement.

References

1. Lou X, Shi G, Chen F, Yang M (2010) Active lateral pressure of loop-cone shaped granular material against silo wall. J. Journal of Geotechnical Engineering S2:25–28 (in Chinese)
2. Janssen, H.A. (1895). Versucheübergetreidedruck in silozellen. *J. ZdesVereinesDeutscherIngenieure*, 39(35), 1045–1049
3. Xiong J, Meng S, Zhou Z (2010) Analysis in wall stress of large-diameter prestressed concrete coal silo under internal pressure. J. J CHINA COAL SOC 02:48–52 (in Chinese)
4. Chen C, Liang X, Shi X, Yang B (2007) Application analysis of earth pressure computation formulae in large squat silos. J. CHIN J ROCK MECH ENG 026(002):421–426 (in Chinese)
5. Gonzalez-Montellano C, Gallego E, Ramirez-Gomez A, Ayuga F (2012) Three-dimensional discrete element models for simulating the filling and emptying of silos: analysis of numerical results. J. Comput. Chem. Eng. 40:22–32
6. Ketterhagen WR, Curtis JS, Wassgren CR, Hancock BC (2009) Predicting the flow mode from hoppers using the discrete element method. J. Powder Technol. 195(1):1–10
7. Wang Y, Lu Y, Ooi JY (2014) Finite element modelling of wall pressures in a cylindrical silo with conical hopper using an arbitrary lagrangian–eulerian formulation. J. Powder Technol. 257:181–190
8. Vidal P, Gallego E, Guaita M, Ayuga F (2008) Finite element analysis under different boundary conditions of the filling of cylindrical steel silos having an eccentric hopper. J. J. Constr. Steel. Res. 64(4):480–492
9. Wang B, Chen M, Tang Z, Yuan Z (2010) Finite element analysis of huge coal silo foundation. J. ROCK SOIL MECH 031(006):1983–1988 (in Chinese)
10. Yang, H., Yang, D., & Zhao, Y. (2011). Three-dimensional finite element simulation of static granular material pressure for steel silos. *J. J Zhejiang Univ.SCI*, (08), 101–107. (in Chinese)
11. Zhang J, Song C, Bai J, Wang Q, Huang C (2007) Design method and computer simulation of large sized coal storage silo in thermal power stations. J. J CHINA COAL SOC 09:989–992 (in Chinese)
12. Wang Y, Lu Y, Ooi JY (2013) Numerical modelling of dynamic pressure and flow in hopper discharge using the arbitrary Lagrangian -Eulerian formulation. J. Eng. Struct. 56:1308–1320
13. Ding S, Rotter JM, Ooi JY, Enstad G (2011) Development of normal pressure and frictional traction along the walls of a steep conical hopper during filling. J. Thin-Walled Struct. 49(10):1246–1250
14. Yang, X. (2011). Research of strength and stability of flat-bottom squat steel silos strengthened with steel sheet strips. Dissertation, Shandong University

Chapter 8
Application of DEM on the Design and Performance Analysis of Bionic Subsoiler

8.1 PFC^{3D} Simulation Model

8.1.1 Black Bear Claw Model

In this chapter, the claw of black bear was selected as the research object (Fig. 8.1), which comes from an adult Ursus americanus (Manitoba, Canada). The black bear is a large bear with a body length of 120–220 cm, a male weight of 60–225 kg and a female weight of 40–150 kg. It can only stand and walk with hind legs, with flat sole and slow walking. The hind legs are slightly longer than the forelegs, with five toes in both front and rear legs. The black bear mainly lives in coniferous forest, deciduous forest and woodland. It has strong adaptability. In the spring after hibernation, it forages on the low slope and move to the high slope in summer. When it is in danger, it will climb to the tree. The black bear hibernates in winter and builds a cave in the tree hole, log or cave ditch. The black bear, with sharp and curved claws, usually lives in bushes, and its claws are its main tool for digging caves. Figure 8.2a is the forepaw of the black bear, with five toes, of which the middle claw toe (Fig. 8.2b) is the largest and most important digging claw toe.

The specific steps are as follows: firstly, the 3D laser scanner (Fig. 8.3) needs to be calibrated to ensure that it can scan accurately, and the point cloud image can be obtained by scanning the various angles of the claw toe by the 3D laser scanner. Then use the reverse modeling software Geomagic studio (version 11, Geomagic, The United States, North Carolina) to register and stitch the point cloud, triangulate the simplified model, simplify and smooth it, and finally get the claw-toe model in STL format. Figure 8.4 shows the claw model acquisition process.

Figure 8.5 shows the scanning middle claw toe structure, which is described by some key points. Point o is the tip of claw toe, and point q is the salient point on

X. Wang et al., *Engineering Applications of Discrete Element Method*, Engineering Applications of Computational Methods 4,
https://doi.org/10.1007/978-981-15-7977-6_8

Fig. 8.1 The black bear

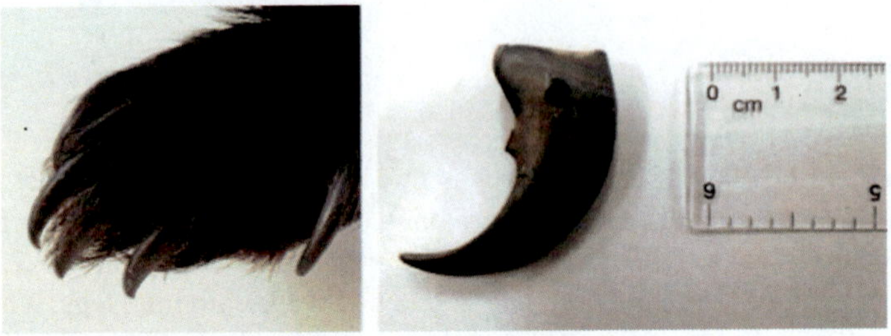

Fig. 8.2 Bear paw and Bear claw

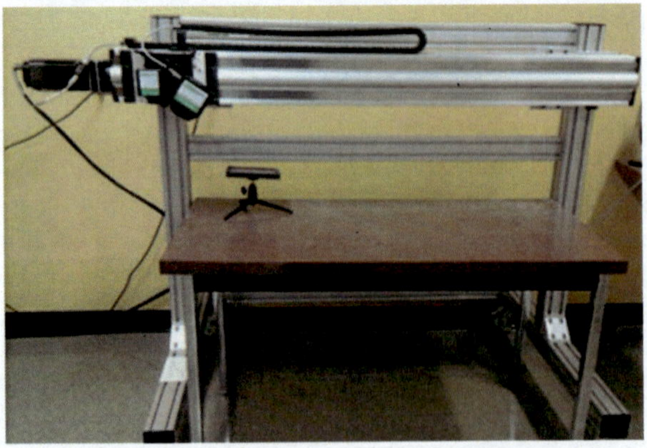

Fig. 8.3 The 3D laser scanner

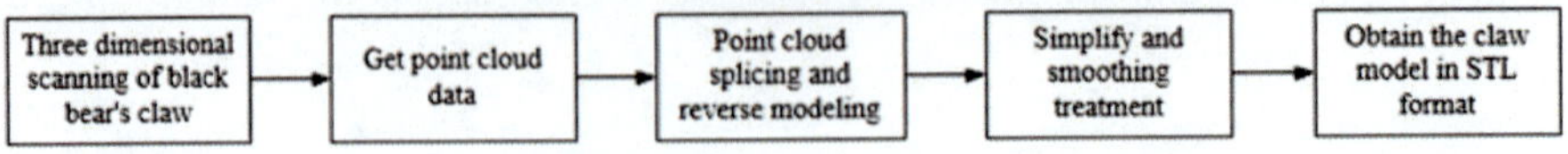

Fig. 8.4 The claw model acquisition process

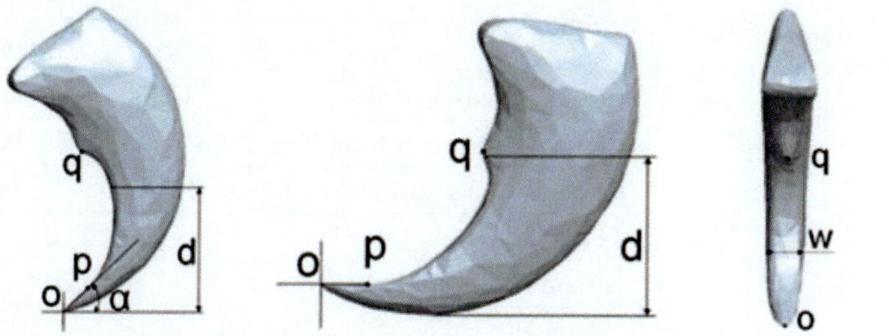

Fig. 8.5 The model claw: op is the linear part of the claw tip; pq is the curve line; a is rake angle; d is working depth; w is working width

claw toe. In nature, the interaction between claw toe and soil occurs below point q, so the loosening depth of this study will not exceed point q; and point p is the tip of the horizontal part of claw. Three working parameters are used to describe the claw: working depth, rake angle and working width. The working depth d is defined as the depth of the toe in the soil, and the angle α is defined as the angle between the straight part op at the front end of the toe and the horizontal line. Figure 8.5a and b show the rake angle of 45° and 0°, respectively. The working width w changes with the depth of the toe in the soil, which is defined as 5.5 mm in this study. The dimension is measured at 1/2 of the working depth under the condition of 25° angle in the soil.

8.1.2 Soil Particle Model

The soil particles used in this simulation were spherical particles with a diameter of 2.4 mm. The particles were generated by the radius expansion method, and the generated area was a box without the roof. When the maximum unbalanced force of particles is less than 0.001 N, it is considered that the particles have reached a static state. Finally, the soil size of area is 120 mm × 100 mm × 35 mm, which makes the claw toe 17 mm away from the bottom of the soil groove and 47 mm away from the side. Such a size can ensure that the claw toe has enough space to run and will not be disturbed by the wall.

The particle model used PFC^{3D} parallel-bond model. The micro-particle parameters required for the simulation are from the research team of Manitoba on sandy loam [2] (Table 8.1). Through the sensitivity analysis of the parameters, it is found that the normal and tangential contact stiffness of particles has the greatest influence on the cultivation resistance and the soil movement form, so these two parameters need to be corrected again.

Table 8.1 Parameters of model particles for a sandy loam soil

Parameter	Value
Particle density/kg/m^3	2650
Particle normal stiffness (k_n)/N/m	To be calibrated
Particle shear stiffness (k_s)/N/m	To be calibrated
Particle friction (μ)	0.5
Bond radius multiplier (r_p)	0.5
Parallel-bond normal stiffness (k_{pn})/Pa/m	1.042×10^{10}
Parallel-bond shear stiffness (k_{ps})/Pa/m	1.042×10^{10}
Parallel-bond normal strength (k_{pnst})/Pa	2×10^4
Parallel-bond shear strength (k_{psst})/Pa	2×10^4
Local damping coefficient (θ)	1.0
Viscous damping coefficient (ω)	0.5

8.2 Soil Parameter Calibration

8.2.1 Calibration Scheme Design

In order to make the correction more convenient, it can be assumed that the two are equal, so only k_n needs to be corrected. However, because the size of the claw is too small and the working resistance is relatively small, the test of the claw will be extremely difficult. Therefore, through the simulation of the interaction process between the soil and the bulldozer plate (Fig. 8.6), the simulation results are compared with the theoretical prediction values, so as to correct the soil stiffness [3].

The width of the bulldozer plate was 8 mm, the simulated running speed was 0.83 m/s (within the range of common tillage speed), and the working depth of the bulldozer plate was set at 18 mm (the same maximum working depth as the toe of the claw) and the rake angle was 20.5° (Fig. 8.6). Through setting different k_n values for simulation, the simulation results of bulldozer plate under different k_n values

Fig. 8.6 The calibration of the model particles

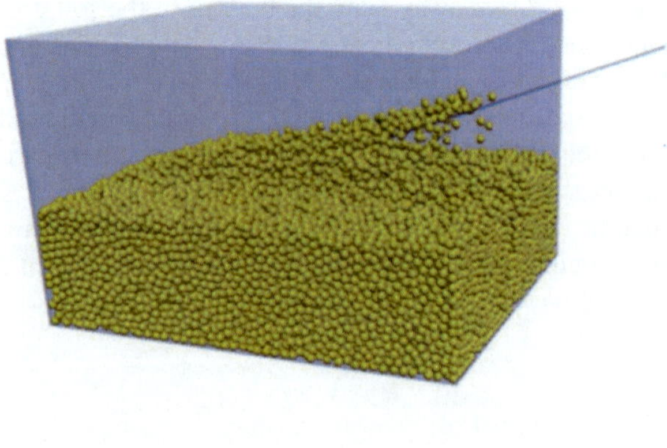

were obtained, and compared with the predicted values of soil motion equation to determine the optimal k_n value.

8.2.2 *Influencing Factors and Measuring Methods*

The rake angle and working depth of the machine are important factors that affect the working effect of the machine. The working effects include soil resistance and soil disturbance [4]. In general, with the increase of the rake angle, if the tillage resistance increases first and then decreases, it means that a certain rake angle will minimize the tillage resistance. However, the working parts vary in structure and the angle of penetration varies [5, 6]. Ucgul et al. [7] studied the influence of shovel rake angle on horizontal resistance and groove profile of non-cohesive soil. Horizontal resistance increases with the increase of rake angle from 10° to 45°, while vertical resistance increases first and then decreases with the increase of rake angle, reaching its maximum value at 20°. Ucgul et al. [8] also studied the effect of tillage depth and speed on horizontal resistance. Depth has a greater influence on horizontal resistance of shovel than speed when soil viscosity is taken into account. Tong and Moyad [5] developed a program based on C ++ to analyze the effect of rake angle on soil failure plane parameters. It was found that the angle of soil failure plane decreases with the increase of soil rake angle from 15° to 75°. Mouazen and Neményi [9] tested and compared four groups of subsoilers with different rake angles. The study found that the dynamic consumption was minimal when the rake angle was 15° and the handle inclination angle was 75°. In this chapter, similar methods are used to analyze the tillage performance of the machine at different rake angles and working depths, and the tillage performance of the machine is evaluated by the comprehensive effect evaluation coefficient.

8.2.2.1 Prediction of Theoretical Values

The equation of soil motion used to predict resistance is:

$$P = \left(\gamma g d^2 N_\gamma + c d N_c\right) w \tag{8.1}$$

where P is the resistance of soil, γ is the specific gravity of soil, g is the acceleration of gravity, d is the working depth of working parts, c is the bonding strength of soil and w is the width of working parts.

Horizontal and vertical resistance can be obtained by:

$$F_d = P \sin(\alpha + \delta) \tag{8.2}$$

$$F_v = P \cos(\alpha + \delta) \tag{8.3}$$

Table 8.2 The parameters used for the universal earthmoving equation

Parameter	Value	From
Soil bulk density (γ)/kg/m^3	1300	Campbell (1985)
Soil bonding strength (c)/Pa	17500	Nandanwar (2015)
Tool-soil friction angle (δ)/°	23.5	Godwin (2007)
Soil friction angle (ϕ)/°	17	Nandanwar (2015)
Impact factors ($N\gamma$)	1.11	Calculated
Impact factors (Nc)	2.07	Calculated
Tool width (w)/m	0.0055	Selected
Tool rake angle (α)/°	20.5	Selected
Working depth (d)/m	0.018	Selected

where F_d is the horizontal resistance; F_v is the vertical resistance; α is the angle of penetration; δ is the angle of friction between the soil and the tool.

N_γ and N_c are parameters determined by the soil friction and the geometry of the working parts, taken from the wedge theory [4], and the remaining parameters are taken from the previous measurement of the same type of sandy loam by our team [10]. The values of all parameters are listed in Table 8.2.

The soil disturbance width is predicted by the following equation:

$$L = w + 2s \tag{8.4}$$

where L is the disturbance width of soil; s is the disturbance width of both sides of the bulldozer plate, and the calculation method is as follows:

$$s = r\left[1 - (d\cot\alpha/r)^2\right]^{1/2} \tag{8.5}$$

where r is the front failure length of soil.

8.2.2.2 Measurement of Simulation Results

(1) Horizontal and vertical resistance

In the process of DEM simulation of the interaction between claw and soil, the horizontal and vertical resistance of claw can be measured. Figure 8.7 shows the horizontal and vertical resistances measured at the rake angle of 25° and the working depth of 18 mm. Due to the random distribution of soil particles, the simulated force measured fluctuates within a certain range. When the claws begin to contact the soil, the horizontal and vertical resistance begin to increase. Then as the claws continue to move in the trough, the force is maintained in a stable state. The average force is taken from 40 mm to 110 mm, and the soil disturbance is also taken from this part.

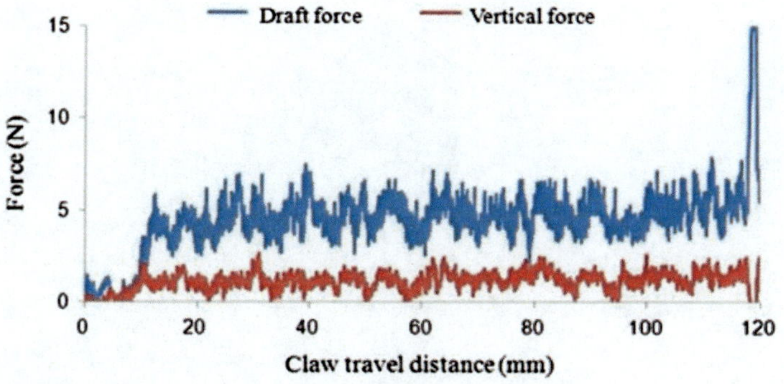

Fig. 8.7 An example of force curves from the simulation

(2) Soil disturbance

Soil disturbances are described in this chapter by changes in soil porosity *u* and discharged soil *n*. *u* is defined as the difference between the final porosity and the initial porosity, which reflects the loosening effect of claw. Discharge soil *n* refers to the number of soil particles discharged above the original soil surface after the toe of the paw passes through. It is related to the lifting action of working machines [11]. Three measuring circles in the PFC^{3D} were used to measure soil porosity. The center of the measuring circle is located at 60, 80 and 100 mm from the starting point along the direction of movement of the claw on the soil surface. Only the lower half of circle is below the soil surface, so all measurements are based on the lower half of circle. The measuring circle radius is 18 mm, under which the maximum working depth of claw toe can be included. The initial porosity has been measured before claw operation (Fig. 8.8a). When the claw toe passes through and the soil particles reach a stable state, measure the number of particles on the initial soil surface (Fig. 8.8b). Next, the particles on the soil surface are removed and the final porosity is measured by using a measuring circle (Fig. 8.8c). The width of soil disturbance is shown in Fig. 8.8d.

(3) Comprehensive evaluation coefficient of cultivation

The main factors that affect tillage performance are tillage resistance, soil discharge and soil porosity. The smaller the horizontal resistance is, the less the soil is discharged, and the larger change of soil porosity means the better tillage performance of tillage tools. For the influence of vertical resistance, there is no definite conclusion at this stage. On the one hand, a certain vertical resistance is convenient for the subsoiler to enter the soil and maintain the working depth; on the other hand, the excessive vertical resistance will cause the excessive sliding friction force of the subsoiling working depth limited wheel. Based on these two reasons, comprehensive evaluation coefficient of tillage does not include vertical resistance. In addition, the comprehensive evaluation coefficient is only suitable for the comparison between two subsoilers with the same tillage depth. Only in this way can the comparison of

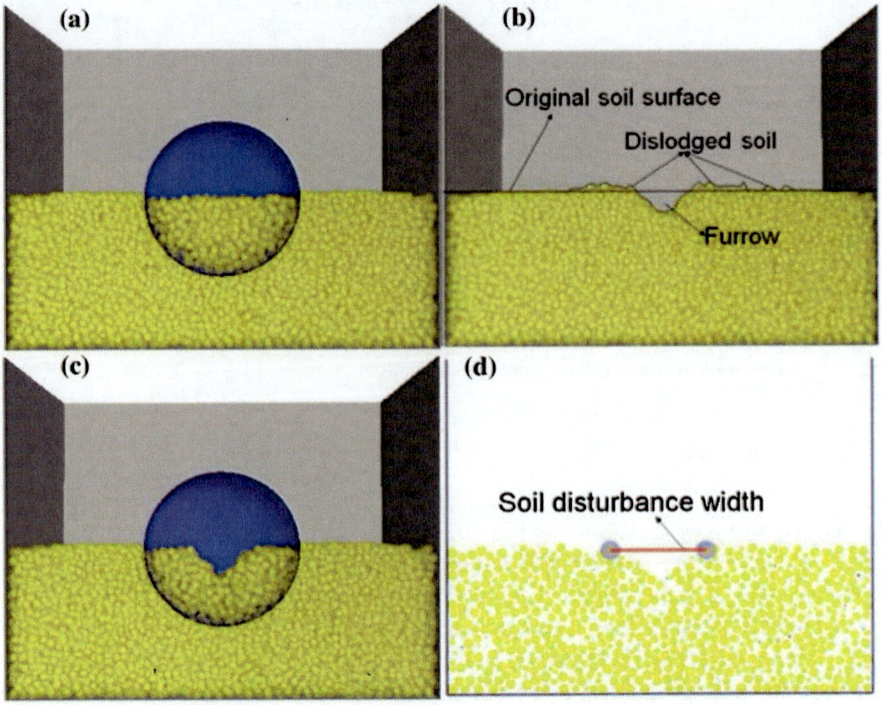

Fig. 8.8 Screenshots of the soil cross-section showing soil disturbance measurements

subsoilers be meaningful. The evaluation coefficient is defined as follows:

$$I = \frac{u}{F_d' \times n'} \tag{8.6}$$

where I is the comprehensive evaluation coefficient of tillage; u is the change of porosity; n' is the ratio of the number of discharged particles per unit length, which means the ratio of the number of discharged particles within the length of 0.1 m to that of 1/0.1 m. The calculation method is as follows:

$$n' = n/1.2 \tag{8.7}$$

F'_d is the horizontal resistance coefficient, which means the ratio of the horizontal resistance F_d to 1000 N, which is calculated as follows:

$$F_d' = F_d/1000N \tag{8.8}$$

(4) Total kinetic energy of soil

Total soil kinetic energy is another variable that describes the dynamic behavior of soil and is directly related to the energy required for soil cutting. The total kinetic energy equation of soil particles during claw movement is as follows:

$$E_k = \frac{1}{2}\sum_{N}\left(m_i V_i^2 + I_i\omega_i \times \omega_i\right) \tag{8.9}$$

where E_k is the total kinetic energy; N is the number of particles; m_i is the inertia mass; V_i is the particle movement speed; I_i is the inertia tensor; ω_i is the particle rotation speed.

8.2.3 Soil Particle Correction Results

The horizontal and vertical resistances of bulldozer plate predicted by soil motion equation are 3.6 N and 3.73 N, respectively, and the width of soil disturbance is 20.9 mm. By comparing the theoretical predicted value with the simulation result, the range of k_n is 1.6 × 10^4 N/m ~ 2.0 × 10^4 N/m. In order to find the simulation value that best matches the three quantities, the calculation method of relative error is as follows:

$$RE = \frac{1}{3}\left(\frac{|D_p - D_s|}{D_p} + \frac{|V_p - V_s|}{V_p} + \frac{|L_p - L_s|}{L_p}\right) \times 100\% \tag{8.10}$$

where RE is the relative error; D_p and D_s are the horizontal resistance of prediction and simulation; V_p and V_s are the vertical resistance of prediction and simulation; L_p and L_s are the soil disturbance width of prediction and simulation.

Table 8.3 shows that the relative error fluctuation of the simulation results is 4.08–10.11% with the change of k_n compared with the predicted results of the soil motion equation. When k_n is 1.8 × 10^4 N/m, the relative error is the smallest (4.08%), which means that when k_n is this value, the simulation results match the theoretical

Table 8.3 Simulation results from the calibration of particle stiffness

Assume (k_n)/N/m	Horizontal resistance/N	Vertical resistance/N	Disturbed width/mm	Relative error/%
1.6 × 10^4	3.54	3.46	24.4	8.52
1.7 × 10^4	3.56	3.39	18.8	6.88
1.8 × 10^4	3.81	3.66	21.8	4.08
1.9 × 10^4	3.82	3.68	19.3	4.99
2.0 × 10^4	3.82	3.65	25.5	10.11

predictions best. Therefore, 1.8×10^4 N/m was selected as the soil k_n value for simulation.

8.3 Analysis of Claw–Soil Interaction

8.3.1 Claw–Soil Interaction Model

The purpose of this chapter is to provide the basis for the optimal design of the subsoiler, and the horizontal movement is the most important movement mode of the farming components in the field. Therefore, this study only focuses on the horizontal movement of the claw in the plane, without considering the spatial movement of the claw.

The claw model was imported into PFC^{3D} and placed at a specified depth and rake angle on the right side of the trough (Fig. 8.9). In order to study the effect of soil rake angle and working depth on soil tillage resistance and soil disturbance, the changing range of claw rake angle is from 0° to 45° every 5°, and the working range of claw is from 3 to 18 mm every 3 mm. To keep the q point on the soil surface and make the curve part of the toe interact with the soil as much as possible, the maximum working depth of the claw was set to 18 mm. When the working depth changes, the rake angle remains constant at 25° and when the rake angle changes, the working depth remains at 18 mm. During all simulation processes, the movement speed of the toe was 0.83 m/s.

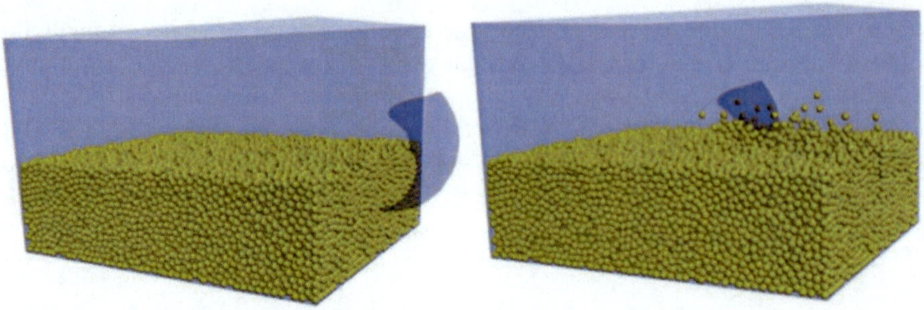

(a) claw at the initial state (b) claw traveling in the soil domain

Fig. 8.9 The model of soil–claw interaction

8.3.2 Claw–Soil Interaction Results

Using the control variable method, only one factor is changed, and the effects of rake angle and working depth on the soil kinetic energy, tillage resistance, discharge soil, and porosity change are analyzed, respectively. When the rake angle changes, the depth remains 18 mm and when the depth of the soil changes, the front angle remains 25°.

8.3.2.1 Effect of Rake Angle and Working Depth on Soil Kinetic Energy

During the interaction between the claw and the soil, the total kinetic energy reflects the total dynamic changes of soil particles in the soil trough. When the rake angle increases from 0° to 45°, the kinetic energy fluctuates very little, from 8.79×10^{-4} to 1×10^{-3} J (Fig. 8.10a). The smaller kinetic energy at the rake angle of 0° is due to the smaller particle flow at this rake angle. When the depth of the soil changes from 3 to 18 mm, the kinetic energy increases as an exponential function from 3.01×10^{-4} to 9.51×10^{-4} J (Fig. 8.10b). The larger the depth of the soil, the more particles will be disturbed, so the kinetic energy will increase accordingly.

For a more intuitive view of the dynamic movement of soil particles during claw toe movement, Fig. 8.11 shows a cross-sectional view along the direction of claw toe movement. The velocity field shows that the disturbance area in front of the claw is different under different angle of penetration. At the rake angle of 0°, more soil particles move above the surface, while for other penetration angles, more particles are pushed forward (Fig. 8.11b and c). A part of the particles under the tip of claw are disturbed, and the disturbed particles gradually decrease with the increase of the rake angle. In the velocity field, the area of the disturbed area and the number of particles change little with the increase of the rake angle, which may be the reason why the overall kinetic energy is relatively stable. The overall kinetic energy in Fig. 8.11.a can be explained in Fig. 8.11 as the working depth increases. Under the condition of constant angle of penetration (25°), when the depth is 3 mm, only a few soil particles

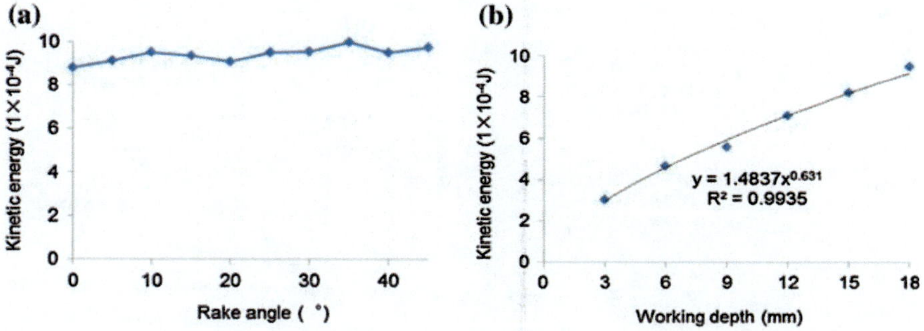

Fig. 8.10 The kinetic energy of soil particles

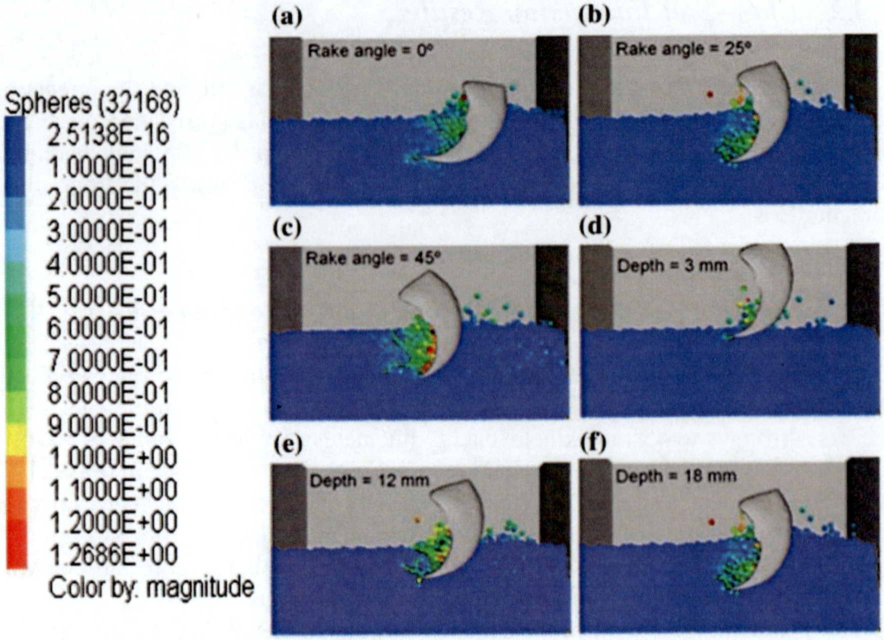

Fig. 8.11 Screenshots of velocity fields; a-c. at different rake angles, depth is 18 mm; d-f. at different depths, rake angle is 45°

are disturbed, and with the increase of operation depth, the number of disturbed soil particles increases.

8.3.2.2 Effect of Rake Angle and Working Depth on Tillage Resistance

As can be seen from Fig. 8.12a, the horizontal resistance is much larger than the

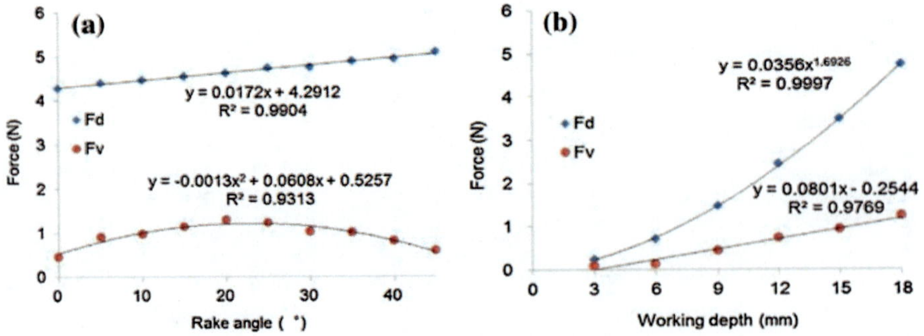

Fig. 8.12 The claw forces, F_d is draft force; F_v is vertical force

vertical resistance, and the average value of the horizontal resistance is about five times of the vertical resistance. With the change of rake angle, the changing trend of horizontal resistance curve is completely different from that of vertical resistance curve. The relationship between horizontal resistance and rake angle is linear equation, and the relationship between vertical resistance and rake angle is quadratic polynomial. Horizontal resistance increases slowly with increase of rake angle. Godwin [12] analyzed the spade tips at different filling angles (22.5–112.5°) and found that the horizontal resistance increased slowly with the increase of rake angle between 22.5° and 45°. When the rake angle reaches 20°, the vertical resistance increases to the peak value (1.30 N), and then decreases gradually with the increase of rake angle. Ucgul [7] has studied the operation process of the shovel and found that the vertical resistance increases first and then decreased with the increase of the rake angle, reaching its maximum value at 20°. The nonlinear growth pattern of the claw of the black bear is related to the complex structural shape of the claw.

When the working depth increases from 3 to 18 mm, the horizontal resistance increases from 0.23 to 4.74 N (Fig. 8.12b). Scholars have studied that the relationship between horizontal resistance and work depth is linear growth [13, 14]. However, in this study, we found that the relationship between horizontal resistance of claw and working depth is nonlinear. This nonlinear relationship can be expressed by power function equation with a correlation coefficient of 0.99. Vertical resistance increases linearly with increase of working depth, but the trend of increase is slower than that of horizontal resistance.

8.3.2.3 Effect of Rake Angle and Working Depth on Discharge Soil

The amount of discharged soil first increases and then decreases with the increase of the rake angle (Fig. 8.13a). A total of 407 soil particles are discharged to the surface of the soil at an angle of 10°, which results in rough soil surface which is unfavorable for seeding. At a rake angle of 25°, as the depth increases, more particles come into contact with the claw, so more particles are disturbed. However, the nonlinear growth trend in Fig. 8.13b shows that the number of soil particles discharged is not directly

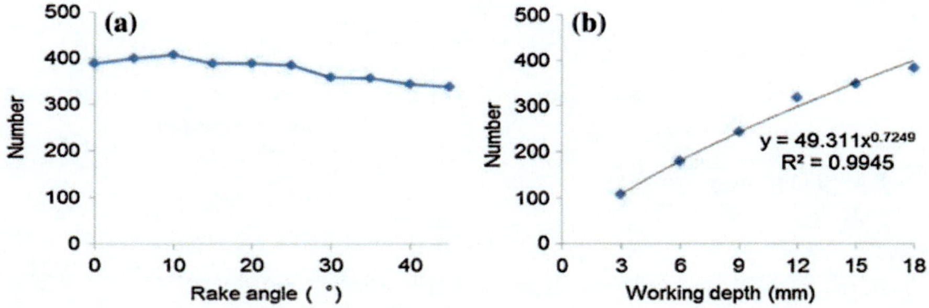

Fig. 8.13 Number of soil particles above the original soil surface

related to the number of disturbed particles. As the depth increases from 3 to 12 mm, the number of discharged particles increases dramatically. Then the growth slows down from 12 and 18 mm. The reason for this trend is that when the depth exceeds 12 mm, a large number of particles begin to fill back into the groove formed by the passage of the claw, resulting in a reduction in the number of particles discharged to the surface of the soil, and the overall change trend of which can be expressed as a power function.

8.3.2.4 Effect of Rake Angle and Working Depth on Porosity Change

The initial porosity of the soil is 0.432 measured by the simulation. From Fig. 8.14a, it is found that the change of porosity has a slight downward trend when the rake angle increases from 0° to 45°. The average change of total porosity is 0.115, which indicates that the effect of loosening is obvious. Porosity variation increases significantly when the operational depth increases from 3 to 18 mm (Fig. 8.14b). The trend can be expressed by power function with correlation coefficient of 0.9973.

In this chapter, the rake angle is defined by the angle between the straight part at the front of the claw and the horizontal direction, which describes only the front of the claw. The shape of the claw has a great influence on soil disturbance and resistance. When the rake angle of the claw changes, the claw also rotates in the vertical plane, which makes the claw shape more horizontal or vertical. In the passive soil failure theory [4], it is pointed out that a more horizontal tool will have a shorter soil failure line and a more vertical tool will have a stronger driving effect than a vertical upgrade effect. This theory can explain the nonlinear relationship between the soil disturbance and the rake angle.

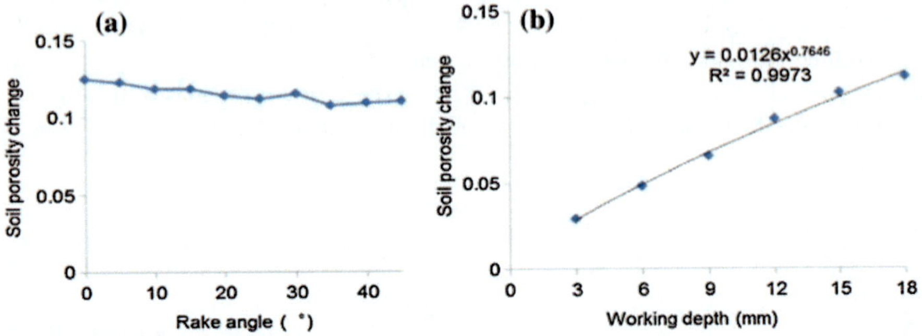

Fig. 8.14 Soil porosity change

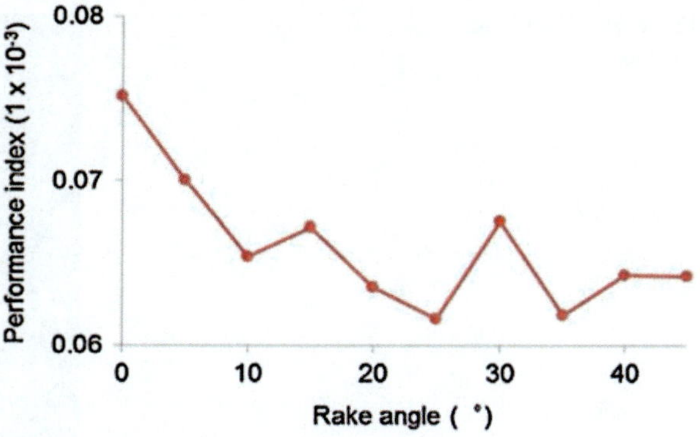

Fig. 8.15 The performance indices for different rake angles

8.3.2.5 Comprehensive Evaluation Coefficient of Tillage

Comprehensive evaluation coefficient of tillage fluctuates from 0.0739 to 0.0902 with the increase of soil entry angle (Fig. 8.15). When the rake angle is 0° and 5°, the values are large. However, neither angle is suitable for the design of the subsoiler. Xu [15] found that when the rake angle is less than 15°, the penetration performance of the subsoiler will be seriously reduced, which results in that the subsoiler cannot meet the operating requirements. The next peak occurs at 30° of rake angle, indicating the highest tillage performance at this rake angle considering both horizontal resistance and soil disturbance.

8.4 Design of Bionic Subsoiler and Study on Subsoiling Effect

8.4.1 Design of Bionic Subsoiler

Using the design method of bionic subsoiler in reference [16] and referring to the structural dimensions of the subsoiler manufactured by CASE Company, the bionic subsoiler based on the claw of black bear paw was designed. 65Mn was used as the manufacturing material. Figure 8.16 shows the subsoiler part of the large-scale combined land preparation machine produced by the company, and the maximum tillage depth of the subsoiler is 450 mm. The frame remains horizontal when the subsoiler is working. And the tilt angle between the upper part of the shovel handle and the frame is 15°. The designed bionic subsoiler also needs to tilt at the same angle during installation. Therefore, when designing the bionic subsoiler, the structure curve is designed with the rake angle of 15°. In this way, when the subsoiler is installed on the frame, its working position is similar to that of the black bear claw. The claw worked at 30° of rake angle.

Fig. 8.16 The CASE subsoiler

The designed subsoiler based on the black bear claw is shown in Fig. 8.17. Firstly, the curve of the black bear claw is extracted point by point with MATLAB, and then the curve equation is fitted in Origin. The fitting equation of the inner guideline as shown in formula (8.11), and the outer guideline is obtained by translating the inner guideline. The width height ratio of the real black bear's claw toe is 0.862. According to the requirement of the plowing depth, the width height ratio (including the tip part) of the subsoiler is 0.894.

$$y = \frac{1}{0.03 - 0.000469233 \cdot x^{0.71482}} \tag{8.11}$$

8.4.2 *PFC^{3D} Simulation Analysis of Subsoiling Effect*

First of all, the strength of the designed bionic handle and shovel tip of subsoiler was checked by FEM, and the results showed that the subsoiler meets the strength requirements. Then, in order to verify the operation effect of the bionic handle, the bionic subsoiler handle and CASE subsoiler handle were introduced into PFC^{3D}. Two bionic shovel tips are installed on two handles, and two subsoilers were simulated and compared under the condition of 350 mm depth and 0.8 m/s speed. The initial positions of the two shovels are shown in Fig. 8.18. The size of the soil trench established by PFC^{3D} is 1.5 m × 1.2 m × 0.5 m. After two subsoil shovels completely

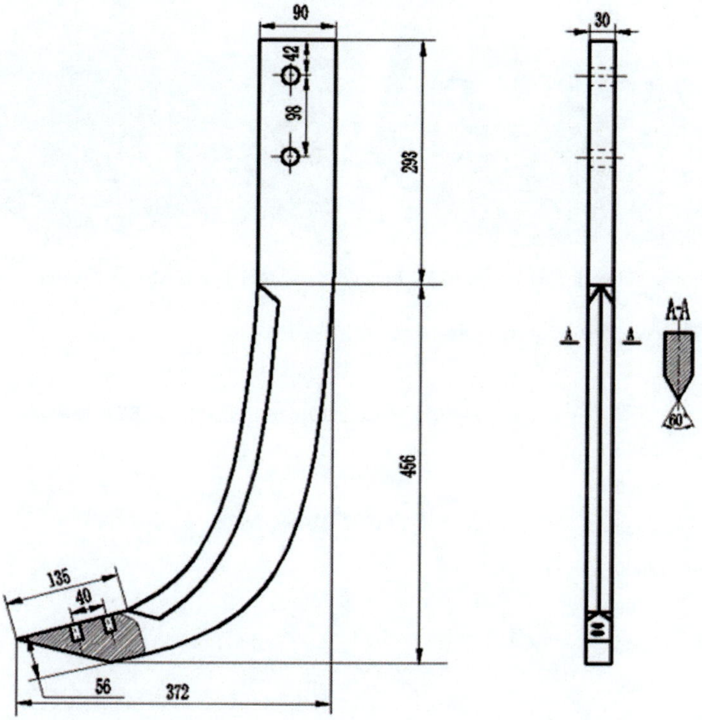

Fig. 8.17 The bionic subsoiler shank based on black bear claw

enter the soil trench, the cultivation resistance is stable. That is to say, when the subsoil shovel runs for 0.5 m, it enters the stable stage. The section with the moving direction of 0.5–1.5 m is taken as the stable section of the subsoiler.

In order to study the designed biomimetic subsoiler tillage performance, this section evaluates these two subsoilers with the comprehensive evaluation coefficient of tillage in formula (8.6), and studies the following items at the stable stage of operation of the subsoiler.

8.4.2.1 Horizontal Resistance

During the operation of the subsoiler, the horizontal resistance suffered during its operation is measured by PFC^{3D}. As shown in Fig. 8.19, the average resistance at the stable stage (0.5–1.5 m) is taken as the horizontal resistance during the operation of the subsoiler.

(a) The bionic subsoiler based on black bear (b) The CASE subsoiler

Fig. 8.18 The comparison of two subsoilers

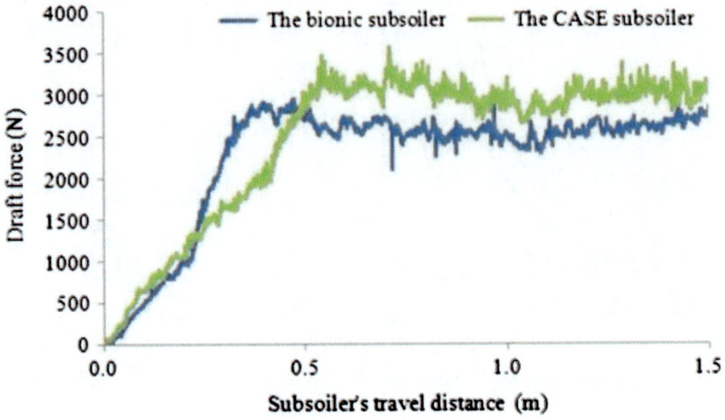

Fig. 8.19 The draft force of two subsoilers

8.4.2.2 Porosity Variation

Porosity is measured by measuring circle in PFC^{3D}. Before subsoiler tillage, the initial porosity of soil in soil trough is 0.4046. After the end of subsoiler tillage, when the particles return to stability, take three measuring circles at the stable stage of subsoiler operation to measure the porosity. The three measuring circles are 0.85, 1 and 1.15 m away from the initial position of the subsoiler, and the measuring circle radius is 0.35 m. The measured porosity after subsoiler tillage is listed in Table 8.4. The average porosity measured by three measuring circles is taken as the porosity after tillage.

Table 8.4 The value of soil porosity after tillage

Subsoiler type	Measurement sphere 1	Measurement sphere 2	Measurement sphere 3	Average soil porosity
Bionic subsoiler based on the black bear claw	0.4066	0.4170	0.4250	0.4162
CASE subsoiler	0.4124	0.4246	0.4350	0.4240

Table 8.5 The tillage performance comparison of two subsoilers

Subsoiler type	Draft force	Soil porosity change	Number of discharged soils	Performance index
Bionic subsoiler based on the black bear claw	2579 N	0.0116	127	4.250×10^{-5}
CASE subsoiler	3239 N	0.0194	1775	4.049×10^{-6}

8.4.2.3 Number of Discharged Soils

A part of soil particles is discharged to the surface during subsoiler tillage, and the amount of discharged soil particles is measured when the particles reach a stable level.

The above results are listed in Table 8.5 and the comprehensive evaluation coefficients of tillage for two kinds of subsoilers are calculated by formulas (8.6), (8.7) and (8.8).

From the results in Table 8.5, it can be found that the horizontal resistance of the subsoiler imitating black bear is less than that of the CASE, which is caused by the curve form of the shank of the deep spade and the presence or absence of the cutting edge. The CASE subsoiler is slightly better at porosity variation, however, the number of particles discharged by the CASE subsoiler is higher than that by the black bear claw subsoiler, which will cause excessive surface roughness. The comprehensive evaluation coefficient of tillage calculated from these three factors shows that the tillage performance of biomimetic subsoiler is 9.5 times that of CASE subsoiler.

References

1. Gupta AK, Jesudas DM, Das PK, Basu K (2004) Performance evaluation of different types of steel for duck foot sweep application. Biosys Eng 88(1):63–74
2. Sadek, M.A., &Chen, Y. (2014). Microproperties calibration of discrete element models for soil-tool interaction. ASABE/CSBE Paper No. 1911030. ASABE, St. Joseph, MI, USA
3. Mak J, Chen Y, Sadek MA (2012) Determining parameters of a discrete element model for soil-tool interaction. Soil & Tillage Research 118:117–122

4. McKyes E (1985) Soil Cutting and Tillage. Elsevier, New York, USA, pp 38–82
5. Tong J, Moayad BZ (2006) Effects of rake angle of chisel plough on soil cutting factors and power requirements: A computer simulation. Soil & Tillage Research 88(1–2):55–64
6. Jafari R, Karparvarfard SH, Hosseini SA (2011) The effect of geometry and motion charcteristics of narrow tillage tool on soil disturbance efficiency. Journal of Agricultural Machinery Science 7(3):253–258
7. Ucgul M, Fielke JM, Saunders C (2014) 3D DEM tillage simulation: Validation of a hysteretic spring (plastic) contact model for a sweep tool operating in a cohesionless soil. Soil & Tillage Research 144:220–227
8. Ucgul M, Fielke JM, Saunders C (2015) Three-dimensional discrete element modelling (DEM) of tillage: Accounting for soil cohesion and adhesion. Biosys Eng 129:298–306
9. Mouazen AM, Nemenyi M, Schwanghart H, Rempfer M (1999) Tillage tool design by the finite element method: part 2. experimental validation of the finite element results with soil bin test. J Agric Eng Res 72:53–58
10. Nandanwar, M. (2015). Simulation of triaxial compression test for sandy loam soil using PFC3D. Dissertation.. University of Manitoba, Winnipeg, Manitoba
11. Salar MR, Esehaghbeygi A, Hemmat A (2013) Soil loosening characteristics of a dual bent blade subsurface tillage implement. Soil & Tillage Research 134:17–24
12. Godwin RJ (2007) A review of the effect of implement geometry on soil failure and implement forces. Soil & Tillage Research 97(2):331–340
13. Bilanski WK, Kushwaha RL (1991) Tillage force prediction from terrain characterization with the bevameter. Soil & Tillage Research 21:299–308
14. Godwin RJ, O'Dogherty MJ (2007) Integrated soil tillage force prediction models. J Terrramech 44(1):3–14
15. Xu, Z. (2009). Design and experimental study of the machine of vibrating deeploose and interillage. Dissertation. Northeast Agricultural University
16. Li B, Liu F, Xia R, Chen J (2015) Distinct element method analysis and experiment of a biomimetic subsoiler. Int. J. Mod. Phys. E 24(1):47–54

Chapter 9
The Effect of Two Wings of Subsoiler on Tillage Based on DEM

Mouazen and Nemenyi compared four subsoilers with different rake angles and observed that the subsoiler with a rake angle of 15° and a shank angle of 75° consumed the least amount of energy [1]. Li et al. found that the best front angle considering the traction and soil disturbance was 30° in studying a bear claw with the objective of designing bionic subsoilers [2]. A program was developed by Tong and Moayad in C++ to analyze the effect of rake angle of a plow on soil failure plane parameters, and the results showed that the angle of soil failure plane decreased with the increase in rake angles from 15° to 75° [3]. Ucgul et al. also simulated the effects of the sweep's rake angle on tillage force and furrow profile of a cohesionless soil and found that draft force increased significantly as the rake angle increased from 10° to 45° [4]. Another study of Ucgul et al. reported that, considering soil cohesion and adhesion, depth had a more pronounced effect on the draft forces than the speed of operation of the sweeps [5]. The subsoiler traction force under different working depths and velocities affected the traction force and fuel consumption of a subsoiling system [6].

The performance of a subsoiling tool also varies with its wing configuration. Typically, a subsoiler consists of a shank and a point. To increase the soil loosening effects of a subsoiler, researchers have added different sized wings to the point [7]. However, it had resulted in an increase in the draft force of the tool. A subsoiler with wide wings increased the draft force by approximately 30%, and doubled the soil disturbance area [8]. The lift height of the subsoiler wings is another parameter that affects the degree of soil fracturing and the size of fractures, which could alleviate subsoil compaction [9]. However, Ahmed and Godwin found no significant differences in subsoiler performance by varying the wing position [10]. A forward dual-bent blade with a rake angle of 7.5° and a bend angle of 10° would be a suitable alternative for the winged chisel plow in shallow soil loosening [11]. These researches demonstrated

X. Wang et al., *Engineering Applications of Discrete Element Method*, Engineering Applications of Computational Methods 4,
https://doi.org/10.1007/978-981-15-7977-6_9

that the effects of rake angle, working depth and wing configuration are tool-specific, so they should be investigated for every new design.

9.1 The Subsoiling Tool and the Soil Parameters in PFC^{3D}

9.1.1 Description of the Subsoiling Tools

The two subsoiling tool design alternatives compared in this chapter were a non-winged subsoiling tool (*NW*) and a winged subsoiling tool (*WW*). The *NW* tool consists of a narrow point and a 30 mm wide edge-on shank (Fig. 9.1a). The point is attached to the shank through two bolts (Fig. 9.1b). The surface of the point and the ground level formed the rake angle of the tool (α). The tractor mounting angle (β) could be adjusted to change the working position of the subsoiling tool, in terms of the rake angle. The penetration angle on the point could also be altered to maintain the working depth during operation. The *WW* tool had the same geometry as the *NW* tool, with an addition of two wings (Fig. 9.1c). The wings were rectangular blades, which were positioned at a 55° angle relative to the side plane of the shank. The main geometrical parameters of the tools are listed in Table 9.1.

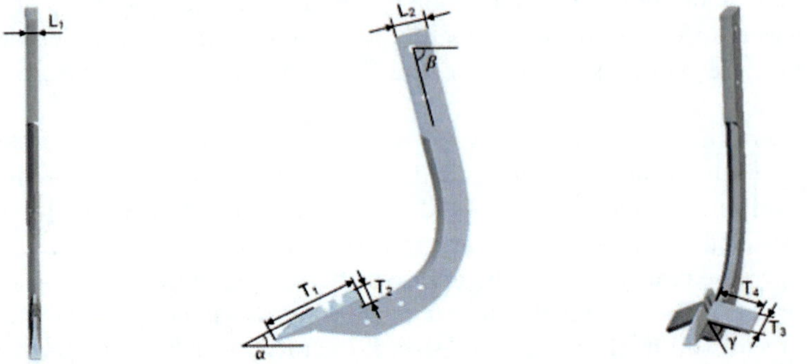

(a) Front-view (b) Side-view of non-winged tool (NW) (c) Winged tool (WW)

Fig. 9.1 Two design alternatives of subsoiling tool

Table 9.1 Geometrical parameters of the two subsoiling tools

Geometrical Parameters	Symbol	Value
Shank width/mm	L_1	30
Shank thickness/mm	L_2	90
Point length/mm	T_1	257
Point thickness/mm	T_2	54
Wing length/mm	T_3	104
Wing width/mm	T_4	135
Wing angle/°	γ	55

9.1.2 Soil Particles and Parameters

A model to simulate the interaction of a subsoiling tool with soil was developed using PFC^{3D}. A simulation tool was implemented with DEM. In the model, soil particles were assumed to be spherical with diameters ranging from 8 to 24 mm, following a uniform distribution (Fig. 9.2a and 9.2b). The soil domain was set as 1.0 m long, 0.5 m wide and 0.5 m high, to avoid the edge effect of the walls and have a practical computing speed, as established by preliminary simulations. The soil domain had a total of 53,634 particles for the simulation. The CAD drawings of the two subsoiling tool designs were converted to STL format files that were integrated into the model particle domain. The contact behavior between the particles was modeled using the parallel-bond model implemented in PFC^{3D} [12]. This contact model has been used by several researchers to successfully model the cohesive behavior of agricultural soils [13–15]. Adding parallel-bonds is similar to adding springs between the particles, and the springs can resist loads applied to the particles. The parallel-bond can break if the external normal or shear force exceeds the perspective bond strength. This phenomenon mimics the process of soil breaking during subsoiling. The particle parameters were adapted from Tamás et al. [14] for a sandy soil with an

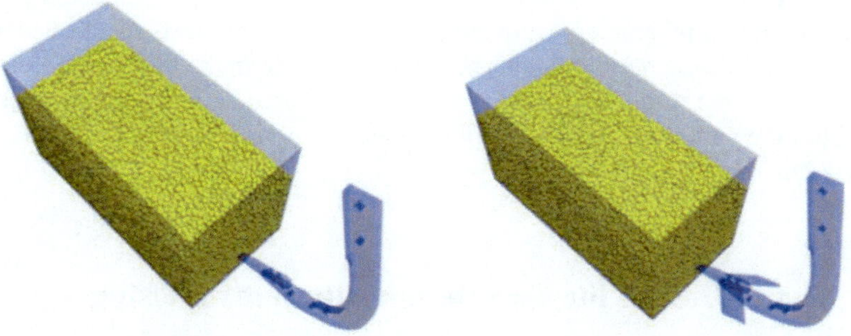

(a) Non-winged tool (NW) (b) Winged (WW) tool

Fig. 9.2 The models of soil–tool interaction

Table 9.2 Parameters of soil particles in the model

Description and unit	Value
Particle normal stiffness/N/m	2×10^6
Particle shear stiffness/N/m	1×10^6
Particle friction	0.6
Radius multiplier	1.0
Normal bond stiffness/Pa/m	3×10^4
Shear bond stiffness/Pa/m	1×10^4
Normal bond strengths/Pa	5×10^3
Shear bond strengths/Pa	1×10^3
Normal/shear critical damping ratio	0.7

initial porosity of 0.42 and bulk density of 1,850 kg/m^3. The values of the parameters are listed in Table 9.2.

9.1.3 Validation of the Model

Draft force data of a standard subsoiling tool were used to validate the model. The data were previously obtained by Li et al. in soil bin tests (Fig. 9.3a) conducted at the Northwest A&F University, Yangling, Shaanxi, China. The soil bin contained a sandy loam soil with moisture content of 19% [2]. The standard tool had an arc-shape and a chisel point (Fig. 9.3.b) which was described in MIS [16]. The tool was run in the soil bin at a constant speed of 0.8 m/s at a depth of 260 mm in three replicates. Draft forces were measured by sensors installed between the soil bin carriage and the three-point hitch of the toolbar. For model validation, the standard subsoiling tool was integrated into the aforementioned model particle domain (Fig. 9.3.c). The stiffness of the tool was assumed to be the same as the stiffness of steel (1×10^9 N/m). The tool was run at a working depth of 260 mm and a travel speed of 0.8 m/s. These operational parameters were the same as the soil bin tests mentioned above. The draft forces were monitored during the simulations. In the model validation, the average draft force simulated by the model was compared with the average measured draft from the soil bin tests. The main geometrical parameters of the standard tool are listed in Table 9.3.

9.2 Virtual Test of the Two Design Alternatives Using PFC^{3D}

After being validated, the model was used to evaluate the performance of the *NW* and *WW* tools. The first set of simulations tested the two subsoiling tools at different rake angles ranging from 23° to 40.5° at a 3.5° interval while keeping the working depth

(a) Setup of the soil bin tests

(b) The model of the soil-standard tool interaction

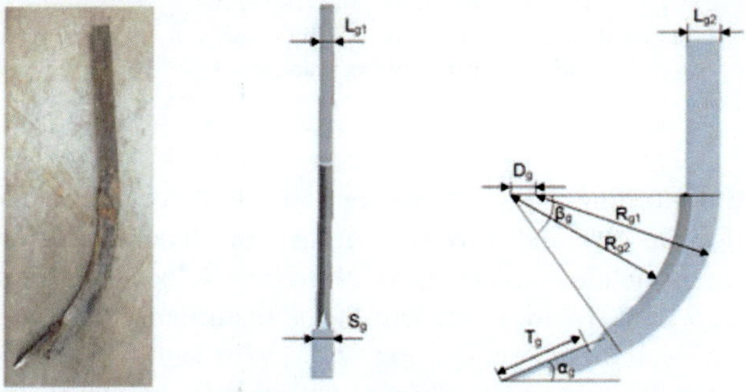

(c) The standard tool

Fig. 9.3 The standard subsoiling tool for model validation

Table 9.3 Geometrical parameters of the standard tool

Geometrical parameters	Value
Shank width (L_{g1})/mm	25
Shank thickness (L_{g2})/mm	60
Point length (T_{g1})/mm	165
Point width (S_g)/mm	40
Outer arc radius (R_{g1})/mm	320
Internal arc radius (R_{g2})/mm	284
Rake angle (α_g)/°	23
Arc angle (β_g)/°	58
Center distance (D_g)/mm	35

constant at 300 mm. The second set of simulations tested the two subsoiling tools at different working depths ranging from 225 mm to 350 mm at a 25 mm interval at the constant rake angle of 30°. All simulations were performed at the travel speed of 0.8 m/s.

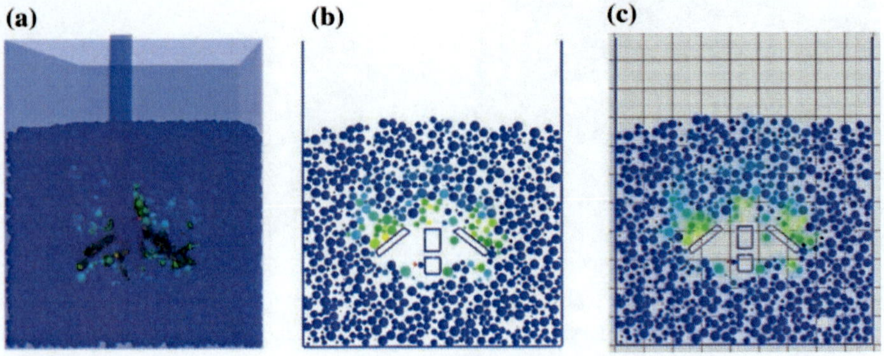

Fig. 9.4 Instantaneous screenshots of the cross-section the model bin: **a** particle velocity contour; **b** a cutting plane showing soil disturbance; **c** the cutting plane overlapped on a grid paper to estimate the soil disturbance area

During the virtual testing, soil disturbance area, draft force and kinetic energy were monitored for the NW and WW subsoiling tools. The soil disturbed can be illustrated using an instantaneous screenshot of particle velocity contours of the soil cross-section (Fig. 9.4a). The measurement of the disturbance area was done on a cutting plane of the soil cross-section (Fig. 9.4b). The cutting plane was created at 0.24 m behind the tip of the tool, which was within the width of the wing. The cutting plane was placed onto an engineering grid paper (Fig. 9.4c) and the number of grids was counted to determine the disturbance area. Three measurements were taken along the soil bin between the travel distance of 0.6 and 0.9 m, which represented the constant speed zone of the model bin.

The draft force was monitored during the simulation, and it was determined to be the total contact force between the particles and the tool in the travel direction. Kinetic energy was also monitored using the built-in function in PFC^{3D}. Kinetic energy is determined by the following equation:

$$E_k = \frac{1}{2}\sum_{N_p}\sum_{i=1}^{6}\mu \tag{9.1}$$

where E_k is kinetic energy; μ_i and v_i are generalized mass and velocity at time steps i, respectively; N_p is the number of particles. The kinetic energy accounts for both translational and rotational motions of all the particles, and reflects dynamics of soil particles and the intensity of tillage.

The draft force of the standard subsoiling tool over the travel distance in the soil domain is shown in Fig. 9.5. The force curve fluctuated over time, which is also the case in practice situation. Over time, the simulated force increased and reached a stable state (Fig. 9.5a). The simulated average draft force of 1,297 N was obtained from the stable section of the model soil bin. A typical curve of the draft forces obtained from soil bin tests is shown in Fig. 9.5b. The force also fluctuated and

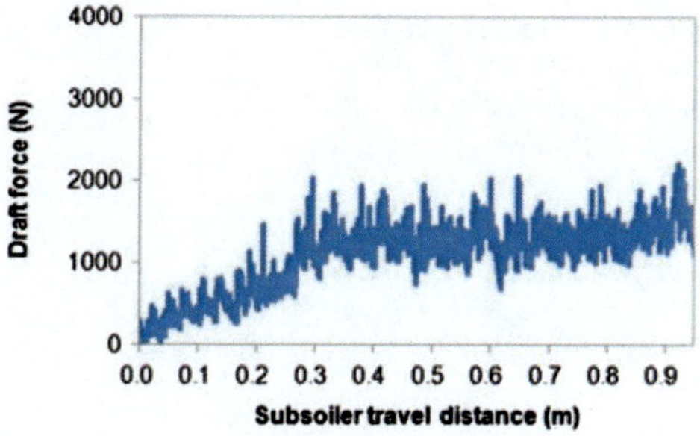

(a) Simulation

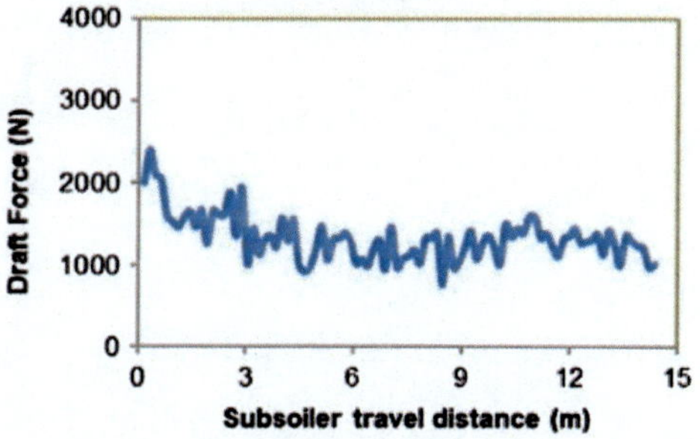

(b) Test

Fig. 9.5 The force–distance curves of the standard subsoiling tool

reached a stable status. The average measured force was 1,342 N. Comparing the average simulated force with the average measured force, one can obtain a relative error of 3.35%. The relative error is small, which shows that the simulation and the measurement can be well matched.

9.2.1 Comparison of the Two Design Alternatives

9.2.1.1 Different Rake Angles

The model was run to predict soil dynamic properties of the *WW* and *NW* subsoiling tools. The results of simulating soil cutting forces and soil disturbance characteristics are shown in Fig. 9.6.

Simulations showed that the draft force of the *NW* tool did not vary much with the rake angle. This was consistent with the findings from Gao et al. [2] for narrow soil-engaging tools. The draft force of the *WW* tool slightly varied with the rake angle; and the lowest draft force (3,614 N) occurred at the rake angle of 26.5° (Fig. 9.6a). Mouazen and Nemenyi found that a subsoiler with a rake angle of 15° and a shank angle of 75° required the least power [1]. In simulation of a sweep, Ucgul et al. obtained a different conclusion: the draft force increased significantly as the rake angle increased from 10° to 45° [4]. The soil disturbance area varied as the rake

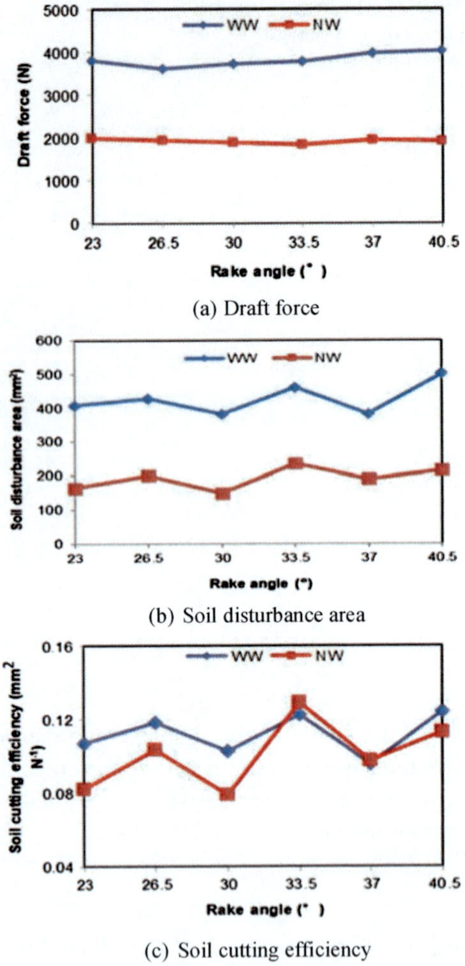

Fig. 9.6 Model results for the winged (WW) and non-winged (NW) subsoiling tools at different rake angles

angle increased, and the same fluctuating pattern was observed for the *NW* and *WW* tools (Fig. 9.6b). The curves had more than one peak, but the overall trend of the disturbance area was increasing with the rake angle.

Spoor and Godwin reported that the subsoiler with wide wings increased the draft force by approximately 30%, and doubled the soil disturbance area [8]. This supports the results of this chapter. Regardless of the rake angle, adding wings to the subsoiling tool doubled the draft force, which was not desirable. However, adding wings also doubled the soil disturbance area, which was advantageous. To evaluate the overall performance of the two designs, a performance index (soil cutting efficiency) was defined as the ratio of soil disturbance area to draft force [17]. At the three smaller rake angles, the *WW* tool had higher cutting efficiency than the *NW* tool; therefore, the wings of the *WW* tool proved to be beneficial at those smaller rake angles. At the three larger rake angles, two subsoiling tools performed similarly, in terms of soil cutting efficiency. For the *NW* tool, the highest cutting efficiency of 0.129 mm^2/N was observed at the rake angle of 33.5°. Therefore, the optimal rake angle of the *NW* tool was 33.5°. Rake angle of 33.5° or 40.5° is recommended for the *WW* tool for high cutting efficiency.

9.2.1.2 Different Working Depths

The effect of working depth was more pronounced and had particular trends when compared with the effect of rake angle. The draft force of the *WW* and *NW* tools increased from 2,776 to 4,486 N and 1,193 to 2,391 N, respectively, as the working depth increased from 225 to 350 mm, both of which showed a linear growth trend, and the correlation coefficient R^2 was greater than 0.99 (Fig. 9.7). These increasing linear trends were consistent with the results reported by Godwin and O'Dogherty [18]. The simulation results showed that adding wings to a subsoiling tool significantly increased the draft force, which was more pronounced at a greater working depth, as demonstrated by a greater slope in the linear regression equation of the *WW* tool at a greater working depth.

The soil disturbance area predicted of the *NW* tool showed a peak at the 325 mm depth (Fig. 9.7b). The reason for this behavior was unknown. The overall trend was constant within the entire depth range of 225–350 mm, with an average soil disturbance area of 175 mm^2. For the *WW* tool, a completely different trend was observed for the disturbance area. The highest soil disturbance area (510 mm^2) was observed at the shallowest depth (225 mm), which is unexpected. As the depth increased, the disturbed area gradually decreased in a nonlinear fashion. These findings indicated that the *WW* tool will have advantages of disturbing more soil when working at shallower depths.

The cutting efficiency of the *WW* tool decreased from 0.1837 to 0.0921 mm^2/N as the depth increased from 225 to 350 mm, which could be described by a power function relationship with an R^2 of 0.94 (Fig. 9.7c). This rapid decreasing trend in cutting efficiency of the *WW* tool was the result of decreasing trend of soil disturbance area and the rapidly increasing trend of the draft force as discussed above. The soil cutting efficiency of the *NW* tool did not show any particular trend. This was due to its highly variable soil disturbance area. The results showed that there would be more advantages of operating the *WW* tool at a smaller depth, in terms of soil cutting efficiency. To minimize the power requirement, both the design alternatives should be operated at the shallowest depth, which enables breaking the hardpan. When comparing the two design alternatives, the addition of wings to the subsoiling tool was advantageous except for the depth of 325 mm.

Fig. 9.7 Model results for WW and NW tools at different working depths

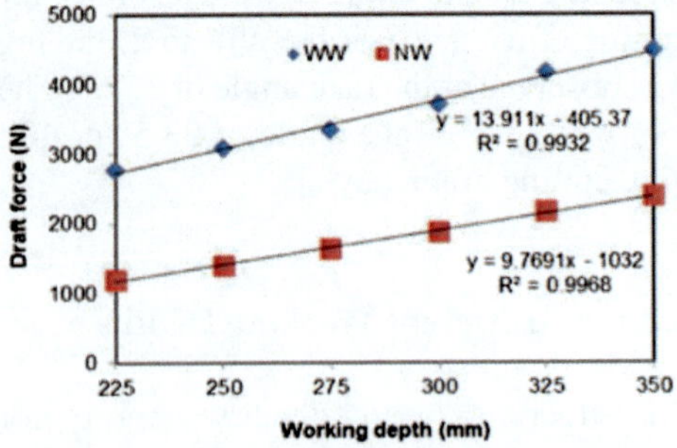

(a) Draft force

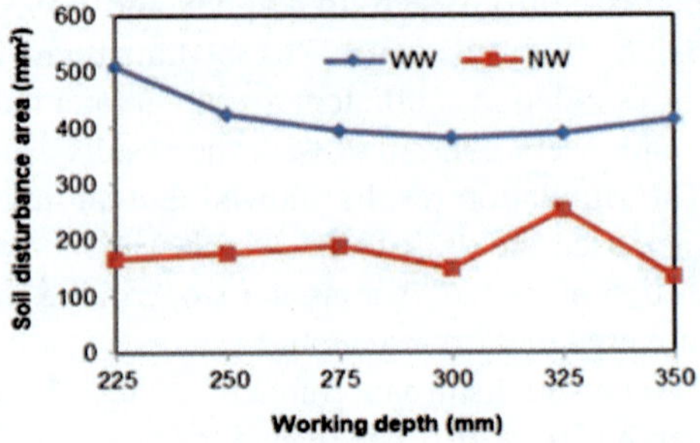

(b) Soil disturbance area

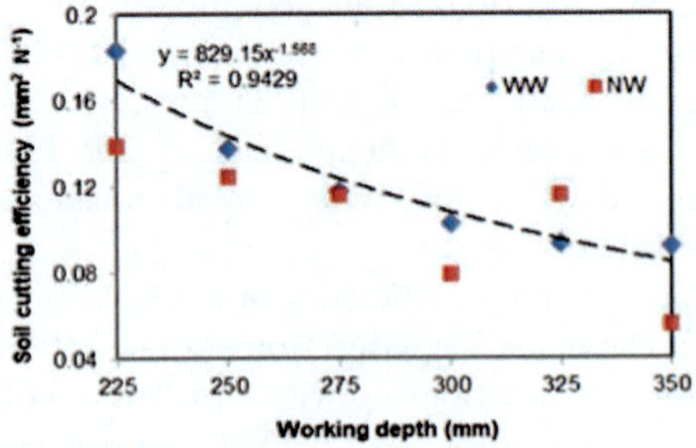

(c) Soil cutting efficiency

9.2.2 *Kinetic Energy of Soil Particles*

For the *NW* tool, the kinetic energy was constant within the range of the rake angles tested (Fig. 9.8a). In other words, the dynamic movement of particles was similar at all rake angles. The kinetic energy of the *WW* tool increased as the rake angle increased. This was due to the stronger lifting and pushing effects of the wings at larger angles. The predicted kinetic energy values of the *WW* tool can be described by a linear relationship with an R^2 value of 0.95. Kinetic energy was proportional to the working depth for both tools (Fig. 9.8b). The relationships were linear over the range of depths examined, as demonstrated by the high values of R^2. The draft

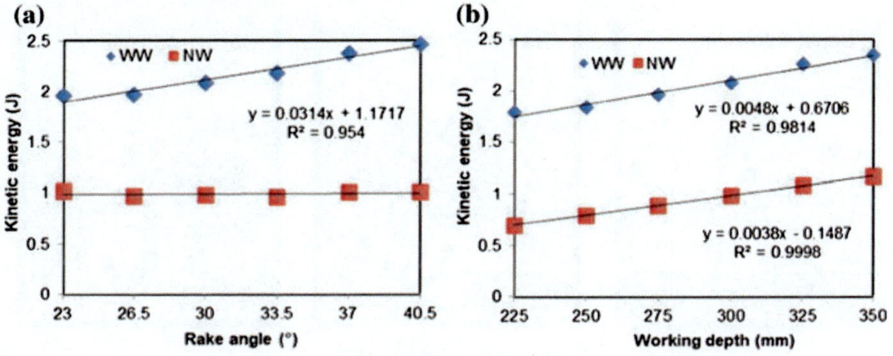

Fig. 9.8 Model results for kinetic energy

forces of the two subsoiling tools were also proportional to the depth (Fig. 9.7a). This implied that higher dynamics of particles reflected higher draft force requirement. Like the other dynamic properties, the kinetic energy of the *WW* tool was higher than the *NW* tool, regardless of the rake angle and working depth.

The kinetic energy results showed that soil particles were more dynamic at higher working depth. This was also reflected by the particles' contact force distribution around the tool, which was demonstrated using the clip boxes features in PFC^{3D}. Figure 9.9 shows some examples of clip boxes located within the wingspans and of the two subsoiling tools at the rake angle of 30°. The contact force distribution below the soil surface was shown at different working depths. A few observations were noticed. There was an existence of void space around both tools, which illustrated the breaking of contact force chains, as a result of tillage action that separated soil particles. This zone of broken force chain corresponded to the predicted soil disturbance area. However, more void space was observed around the *WW* tool than the *NW* tool, meaning more soil particles were impacted by the *WW* tool. This observation supported the previous results that the *WW* tool had larger soil disturbance area and force requirement. The occurrence of more concentrated force chains on the wings of the *WW* tool was also observed, which substantiated lifting and pushing of particles by the wings. The contact force was more concentrated at a greater working depth, especially for the *WW* tool. These observed phenomena are important to the understanding of soil dynamics and cannot be visualized from physical testing of tools. The observations can also be used to explain the results of draft force and soil disturbance. Further examination of soil dynamics at particle level is recommended for future studies.

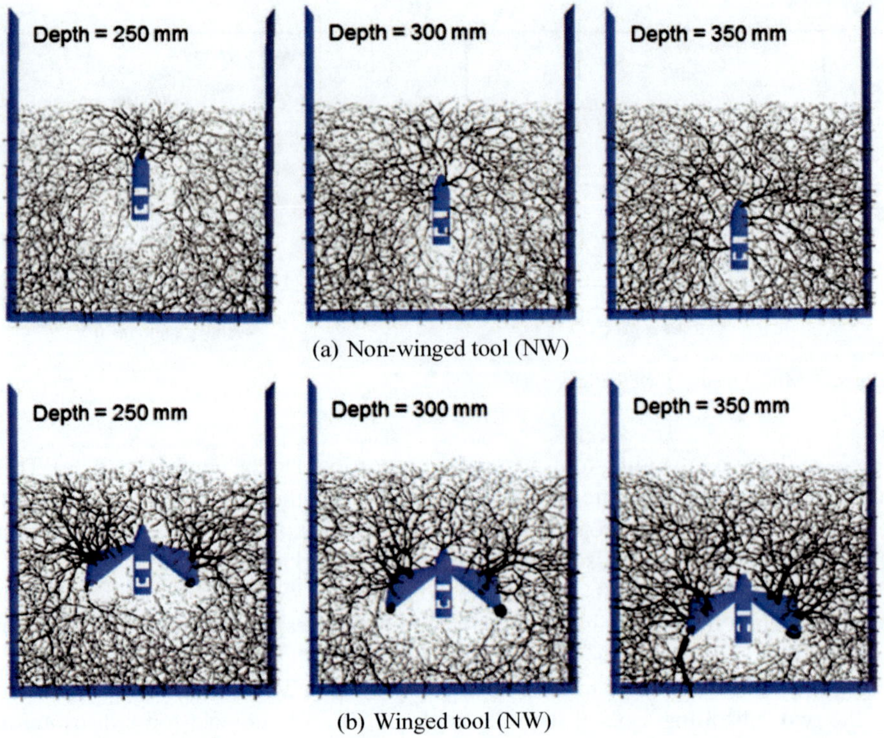

Fig. 9.9 Contact force chain at different working depths

References

1. Mouazen AM, Nemenyi M (1999) Tillage tool design by the finite element method: Part 1. Finite element modelling of soil plastic behaviour. J. Agric. Eng. Res., 72(1), 37–51
2. Li B, Chen Y, Chen J (2016) Modeling soil-claw interaction using the discrete element method (DEM). Soil Tillage Res. 158:177–185
3. Tong J, Moayad BZ (2006) Effects of rake angle of chisel plough on soil cutting factors and power requirements: A computer simulation. Soil Tillage Res. 88(1):55–64
4. Ucgul M, Fielke JM, Saunders C (2014) 3D DEM tillage simulation: Validation of a hysteretic spring (plastic) contact model for a sweep tool operating in a cohesionless soil. Soil Tillage Res. 144:220–227
5. Ucgul M, Fielke JM, Saunders C (2015) Three-dimensional discrete element modelling (DEM) of tillage: Accounting for soil cohesion and adhesion. Biosyst Eng 129:298–306
6. Croitoru, S., Vladut, V., Marin, E., Matache, M.,&Dumitru, I. (2016).Engineering for rural development determination of subsoiler traction force influenced by different working depth and velocity.15th International Scientific Conference on Engineering for Rural Development. Jelgava, Latvia
7. Askari, M., Shahgholi, G., Abbaspour-Gilandeh, Y., &Tash-Shamsabadi, H. (2016).The effect of new wings on subsoiler performance. *Appl. Eng. Agric.*, 32(3), 353–362
8. Spoor G, Godwin RJ (1978) An experimental investigation into the deep loosening of soil by rigid tines. J Agric Eng Res 23(3):243–258

9. Spoor G, Tijink FGJ, Weisskopf P (2003) Subsoil compaction: risk, avoidance, identification and alleviation. Soil Tillage Res. 73(1):175–182
10. Ahmed MH, Godwin RJ (1983) The influence of wing position on subsoiler penetration and soil disturbance. J Agric Eng Res 28(5):489–492
11. Salar MR, Esehaghbeygi A, Hemmat A (2013) Soil loosening characteristics of a dual bent blade subsurface tillage implement. Soil Tillage Res. 134:17–24
12. Itasca. (2015). Particle Flow Code in 3 Dimensions (PFC^{3D}), Theory and Background, version 5.0. Minneapolis, MN: Itasca Consulting Group, Inc
13. Van der Linde, J. (2007). Discrete Element Modeling of a Vibratory Subsoiler. M.S. Dissertation. University of Stellenbosch
14. Tamás K, Jóri IJ, Mouazen AM (2013) Modelling soil-sweep interaction with discrete element method. Soil Tillage Res. 134:223–231
15. Mak, J.,&Chen, Y. (2014). Simulation of Draft Forces of a Sweep in a Loamy Sand Soil Using the Discrete Element Method. *Can. Biosyst. Eng.*, 56(1), 2.1–2.7
16. MIS (1999) JB/T 9788-1999. Machinery Industry Standard of the People's Republic of China, Subsoilerand Share Shaft (in Chinese)
17. McKyes E (1985) Soil Cutting and Tillage. Elsevier Science, New York, NY
18. Godwin RJ, O'Dogherty MJ (2007) Integrated soil tillage force prediction models. J. Terramech. 44(1):3–14

Chapter 10
Conclusion and Prospect

10.1 Work Summary

This book used EDEM to simulate and analyze the scraper conveyor (Chaps. 3, 4 and 5). Chapters 3 and 4, respectively, took coal bulk and scraper conveyor as research objects to simulate the movement of coal bulk under different conditions in the middle trough of scraper conveyor, and obtained the transport state and mechanical effect of coal bulk under different conditions. In Chap. 5, biomimetic middle plate was designed according to the research on non-smooth surface of pit, the optimum wear-resistant structure parameters were obtained, and the wear-resistant mechanism of biomimetic pit-type middle plate was studied.

This book used EDEM to simulate and analyze the coal bunker (Chaps. 6 and 7). In Chap. 6, the influence of different factors on the flow state of coal bulk in coal bunker was studied, and the unloading process of three kinds of hoppers was analyzed. In Chap. 7, the pressure distribution in the silo was analyzed theoretically, the static and dynamic side pressure of the silo was simulated, the side pressure test of the rectangular silo was carried out in the laboratory and the analysis results were verified.

This book used PFC^{3D} to simulate and analyze the subsoiler (Chaps. 8 and 9). In Chap. 8, the model of interaction between the claw and the soil was established. The single-factor variable control method was adopted to study the influence of the angle of penetration and the working depth on the cultivation performance. Finally, the bionic subsoiling handle was designed according to the claw toe curve of the black bear when the claw toe was 30° and compared with the case subsoiling handle. In Chap. 9, the model of interaction between subsoiler and soil was established, and the differences of tillage resistance and soil disturbance range between subsoiler with wing and subsoiler without wing were analyzed.

X. Wang et al., *Engineering Applications of Discrete Element Method*, Engineering Applications of Computational Methods 4,
https://doi.org/10.1007/978-981-15-7977-6_10

10.2 Main Conclusions

(1) Under the stable transport state, the particle size distribution of coal bulk is basically independent of parameters such as mining height, feed speed, transport angle and transport speed, showing the distribution of large pieces on the top and small pieces on the bottom; during the transport process, coal bulk shows the distribution of large particle velocity in the middle part of the middle trough and small particle velocity on both sides.

(2) Under the condition of uphill and downhill, with the angle of uphill and downhill from large to small, the coal between the two scrapers in Y direction flows relatively to the area close to the previous scraper. The highest point of coal accumulation is opposite that it closes to the rear scraper. The distribution of coal between the two scrapers is more uniform and more coal can be transported. Under the condition that the middle trough inclines in the direction of bedding, the coal is concentrated in the trough on the inclined side, and the quality difference of the coal in the trough area on both sides of the chain increases with the increase of the inclination angle. The uneven distribution of coal load on both sides results in weakened synchronism of the tension of the two chains, which is significantly uneven. At the same time, the synchronism is enhanced near the sprocket at both ends and weakened away from the sprocket at the middle. Under different transportation conditions, with the increase of transport volume, the accumulation shape of coal in the middle trough changes from wave-like to stepped, and the increase of effective transport volume of coal requires a process. When the transport rate is less than the loading rate, the accumulation shape of coal gradually increases until both are equal and no longer increases. Under the condition of local accumulation of groove side, there is a sudden change in compression pressure of bulk material in groove side at coal accumulation, and the periodic peak value of compression force of coal material increases and then decreases gradually.

(3) In the scope of study, the wear amount of the middle trough increases with the increase of Poisson's ratio, shear modulus, density and hardness. The wear constant has the most significant influence on the wear amount of the middle trough sample, the shear modulus is more significant, and the Poisson's ratio has the least influence on the density. The wear amount increases with the increase of speed and particle size; the speed mainly affects the normal load by affecting the impact produced when coal particles contact the middle trough, and which in turn affects the wear of the middle trough. The particle size affects the wear of the middle trough by affecting the particle quality and the formation probability of the three-body wear.

(4) For the pit plate under specific working conditions (coal particle size is 6–8 mm, load is 20 N, scraper chain speed is 0.65 m/s and test duration is 6600 s), the order of significant influence is radial distance > depth/diameter ratio > diameter > pitch angle. By comparing and validating the regression prediction model of wear volume and factors, it has been found that the relative error between the

prediction model and the real test is 3.3%. When the ratio of depth to diameter is 1.41, the diameter is 0.69 mm, the pitch angle is 6.55° and the radial distance is 4.66 mm, and the wear resistance of bionic pit plate is improved by 12.6%.

(5) The change of density of coal bulk has no obvious influence on the flow state of bulk, but the larger the particle density, the faster is the unloading speed. The lower the moisture content, the closer is the unloading process to the overall flow, and the faster is the unloading speed. The distribution of particle size has no effect on the flow state of particles. For the parameters set in the paper, the overall flow also appears. The larger the distribution of particle size, the faster the unloading speed. The internal and external friction coefficients between the coal bulk materials have a great influence on the flow state, which shows that the smaller the internal and external friction coefficient, the more stable and smooth the unloading, and the closer it is to the overall flow, the faster the unloading speed. The larger the inclination of the hopper, the closer the flow state of the bulk material to the overall flow, and the faster the unloading speed. The larger the opening size of the discharge hopper, the closer the particles are to the overall flow and the faster is the discharge speed.

(6) Compared with the traditional conical hopper or square hopper, the hyperbolic hopper is easier to unload. Compared with the conical hopper, the unloading time of the hyperbolic hopper is saved nearly 1/3, but the hopper flow form of the particles in the silo is more obvious and the friction to the silo wall is less. When loading, the collision between the particles in the hyperbolic hopper and the silo wall is more serious, and the conical hopper is the most suitable for loading under the same conditions. The confluence of the warehouse wall of the square hopper is easy to form suspended material.

(7) When the coal bunker is static, the lateral pressure distribution and change of the coal bulk material acting on the bunker wall are basically consistent with Janssen's classic calculation theory. As the depth of the straight silo increases, the side pressure gradually increases, and the maximum value appears at the transition position of the straight silo and the hopper silo; while the depth of the hopper silo increases, the side pressure gradually decreases. During the unloading process of the coal bunker, there is a certain degree of increase in the lateral pressure of the bunker wall at each position. During the unloading process, unstable factors increase, the dynamic load on the silo wall increases, and the side pressure has certain fluctuations. The traditional pressure calculation formula is no longer applicable. When the coal bunker is unloaded, the bunker wall is easily damaged by the strength structure, and the impact of various overpressure conditions must be considered in the design.

(8) With the increase of working depth of black bear claw, the total soil kinetic energy, horizontal and vertical tillage resistance, the amount of discharged soil particles and the change of porosity will increase significantly. The soil raking angle has little influence on the total soil kinetic energy, on the amount of discharged soil particles and on the change of porosity. However, with the increase of soil raking angle from 0° to 45°, the horizontal resistance shows a linear growth trend and the vertical resistance has a quadratic function changing

trend. The tillage performance of black bear claw is the best when the rake angle is 30°, and the tillage performance of biomimetic subsoiler is 9.5 times higher than that of CASE subsoiler when the same shovel tip is installed.

(9) Both design schemes with and without wings should be carried out at the shallowest depth to improve soil cutting efficiency. Concerning soil cutting efficiency, the non-winged subsoiling tool shows optimum performance at the rake angle of 33.5°, and the recommended rake angle for the winged tool is 33.5°–40.5°. Adding wings to the subsoiling tool doubles the horizontal resistance and soil disturbance area, regardless of rake angle and working depth.

10.3 Prospect

(1) Application programming interface (API) of EDEM is used to compile relevant program plug-ins to realize functions of ideal particle model generation, real particle shape and contact model improvement. Further improve the coal particle, soil particle model and contact model studied in this book, and build a more consistent discrete element model with the actual situation, so as to make the simulation results more persuasive.

(2) A variety of particle micro-parameter calibration methods are used to calibrate parameters in this book, but there is no uniform conclusion on the method of parameter calibration. Therefore, the research on particle correction methods under different simulation conditions and the accuracy of different correction methods will be the focus of e future research of DEM.

(3) At present, due to the limitation of computer performance, the calculation efficiency will be greatly reduced when the number of particles exceeds a certain range. In the future, with the development and progress of computer level, the calculation of a large number of particles will make the simulation closer to the reality. The computational efficiency will be further improved.

(4) In exploring the wear of coal bulk conveyed by scraper conveyor, due to the function of analysis software, only the wear of coal bulk on the middle trough can be studied, and the wear between scraper chain and the middle trough cannot be considered. Although some scholars have explored the latter, they are not deep and perfect. Therefore, they will try to combine the two wear conditions in the future to further enrich and study the wear theory of middle trough.

(5) Considering the shortage of finite element simulation technology, when analyzing the middle trough of biomimetic pit, the collection of abrasive debris from the morphology of the pit, the processing hardening of the edge of the pit and the lubrication of coal powder cannot be shown in the simulation analysis. In future research, scholars hope to further explore through abrasive wear test and field industrial test.

(6) The bear claw curve is used for the structural optimization design of subsoiler in this book, and has achieved good performance in reducing drag and improving farming effect. There is still a lack of further research on other animals good

at digging caves, and more digging animals' claws need to be studied and compared. And only one curve type subsoilers has been compared with wingless one in this book, and no further analysis has been made on the installation position of various shape subsoilers and wings. Therefore, more detailed analysis is needed in this respect.

(7) This book can effectively promote the application of DEM in engineering, analyze the operation status, fault mechanism and operation effect of relevant equipment in operation, and provide theoretical basis for the optimal design of equipment.